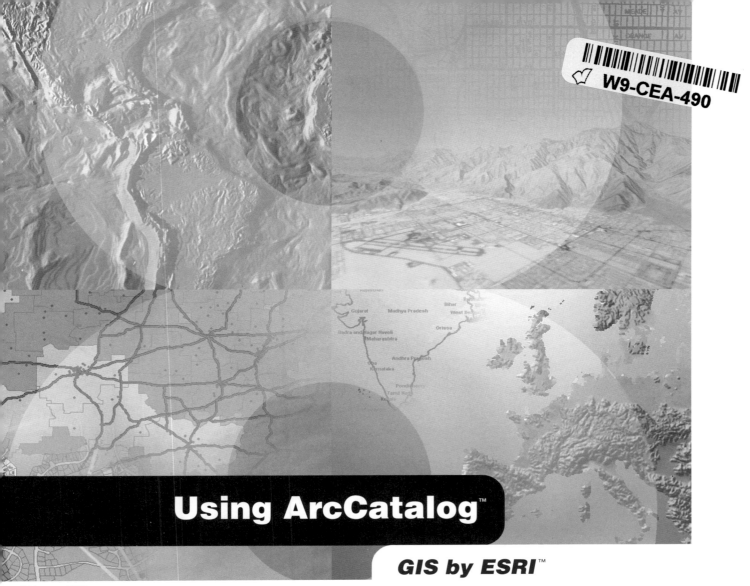

Using ArcCatalog™

GIS by ESRI™

Aleta Vienneau

DATA CREDITS

Geographic data used in the Quick-start tutorial provided courtesy of Yellowstone National Park, National Park Service and is used herein with permission.

Some of the illustrations in this work were made from data supplied by ArcUSA™; ArcWorld™; ArcScene™; ArcEurope™ and Data Solutions; the United States Geological Survey; Space Imaging; the Texas Orthoimagery Program (TOP) and VARGIS; and Geographic Data Technology, Inc., of Lebanon, New Hampshire, and are used herein with permission. Copyright © 2001 Geographic Data Technology, Inc. All rights reserved.

CONTRIBUTING AUTHOR

Jonathan Bailey

U.S. GOVERNMENT RESTRICTED/LIMITED RIGHTS

Contents

Introducing ArcCatalog

1

ESRI® ArcCatalog™ software makes accessing and managing geographic data easy. First, you add connections to the geographic data you work with to the Catalog. You can connect to folders on local disks, to shared folders and databases that are available on the network, or to Internet map servers.

After building your Catalog, you can search for the data you need and explore the search results using the different views that are available. In ArcCatalog, you can work with all data the same way regardless of the format in which it is stored. Several tools are also available to help you organize and maintain your data. For a cartographer, it's never been easier to ensure you're adding the right data to your maps. And, whether you're an analyst managing your personal data holdings or an administrator managing those of a large organization, ArcCatalog simplifies the job.

What can you do with ArcCatalog?

After connecting to a folder, database, or Internet server, you can browse through its contents with the Catalog. You can look for the map you want to print, draw a coverage or page using the values in a table, or find out which coordinate system a raster uses or read about why it was created. Accessing and using information in ArcCatalog is a simple process.

When you've found the data you want to use, add it to a map in ArcMap™ or analyze it using the tools in ArcToolbox™. You might find data that you don't need any longer or that must be altered. The Catalog makes it easy to reorganize your data and modify its properties.

Browse for maps and data

Select a folder, database, or Internet server in the Catalog tree, then examine the list of geographic data it contains in the Contents tab. You can change the appearance of the Contents list using buttons on the Standard toolbar. Switch your view from large icons to small icons, or list properties and metadata for each item to help you decide which is the right one to use. Thumbnail view lists snapshots illustrating the geographic data contained in each item in the folder, database, or Internet server.

Explore the data

Thumbnails give you a quick look at the contents of an item, but you may want to check whether or not a specific feature in a coverage has been updated. Select an item that contains data in the Catalog tree and then examine its data using the Preview tab.

Look at geographic data with Geography view. Buttons and tools on the Geography toolbar let you zoom and pan to explore the features in a computer-aided design (CAD) drawing, cells in a

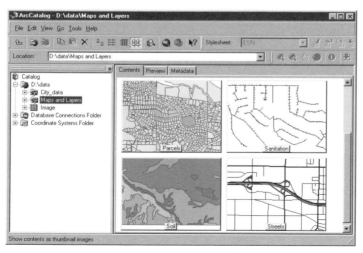

Thumbnails provide an overview of data in the selected folder.

Explore an image before adding it to a map.

raster, or triangles in a triangulated irregular network (TIN). The Identify tool lets you click a feature, cell, or triangle and see its attributes.

With Table view, you can see the attributes of a geographic data source or the contents of any table in a database such as inventory or billing data. Explore the table's contents by rearranging its columns, sorting its rows using the values in one or more columns, or searching for specific values.

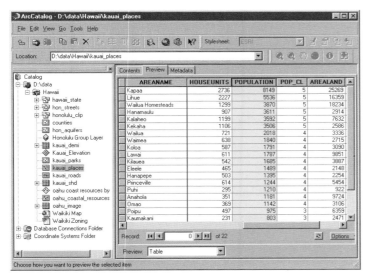

Browse through a shapefile's attributes.

View and create metadata

Before deciding to use a data source in a map, you may need more information. In the Properties dialog box for a data source, you can find its coordinate system and the data type of each attribute. However, for information such as why the data was

created, the appropriate scale at which to use the data, a data accuracy report, or a description of what an attribute name means, you must look at its metadata.

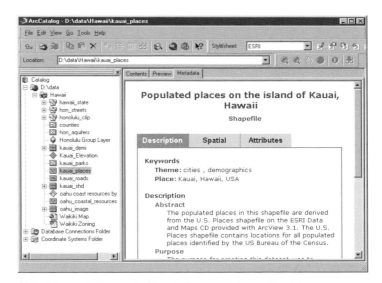

Metadata can help you decide whether or not to use the data.

ArcCatalog comes with a metadata editor that you can use to document your data. The Catalog will fill in as much information as it can using the data's properties. When the data changes—for example, when a new attribute has been added—the next time you look at the metadata the Catalog automatically updates it with the new information. Metadata is an integral part of the data and will follow when the data is copied or moved to a new location.

Search for maps and data

If you know something about the data you are looking for, but you don't know where it is located, ArcCatalog can help. The Search tool will look on disks, in databases, and on ArcIMS™ Internet servers for data that satisfies your criteria.

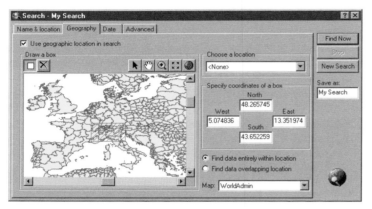

Search for data within your area of interest.

You can search for specific data formats and for maps covering a defined geographic area. You might look for rasters that were published before a given date and that have less than 10 percent cloud cover. Your search is saved in the Catalog. As data is found that satisfies your criteria, shortcuts to those data sources are added to the search's list of results. Later, you can modify the search's criteria and run the search again.

The Search tool uses metadata to evaluate whether a data source satisfies your criteria. Having excellent documentation will be essential for people to be able to find the data they need using the Search tool. This is when metadata is especially important.

Use data in ArcMap and ArcToolbox

After finding the map you want to use, double-click it to open it in ArcMap. Add data to the map by dragging it from the Catalog—for example, dragging it from your search results and dropping it on the map. In addition to creating printed maps, ArcMap is where you go to edit geographic and tabular data.

Instead of adding the data to a map, you may want to convert the data you've found to a different format or buffer its contents. After opening a geoprocessing tool or wizard in ArcToolbox, you can drag the data source from the Catalog and drop it onto the tool. The Toolbox automatically fills in as many options in the tool as possible. If you use a group of tools frequently, you can customize ArcCatalog by creating a new toolbar and adding these tools to it for quick access.

Manage data sources

After looking at the contents of a data source and reviewing its metadata, you might want to modify it to better suit your needs. You can manage the structure of a data source using the Properties dialog box. For example, the Properties dialog box can be

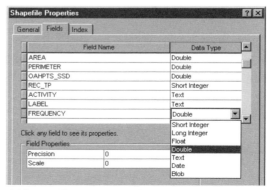

Add an attribute to a shapefile.

used to define a shapefile's coordinate system, generate a coverage's topology, or add an attribute to a table. You can also create a relationship class that defines the relationships between the features in coverages and attributes in INFO™ tables.

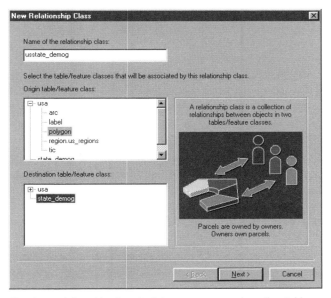

Create a relationship class to link a coverage and another table.

For information on how to use tools in ArcCatalog to design and create new objects in a geodatabase, see *Building a Geodatabase*. If this book is not included with your software package, you'll find these tasks in the 'Working with geodatabases' section in the online Help system's table of contents.

The Catalog makes it easy to organize your data. Delete a coverage by pressing the Delete key on your keyboard. Rename shapefiles and copy tables to another geodatabase just as you would rename and copy files with Windows® Explorer. ArcCatalog makes it easy to start consolidating your spatial data into a coherent library that's distributed across the network.

If you're a data administrator, the Catalog can help you create an environment in which geographic data can be easily used by everyone in your organization. Layers include a shortcut to the data and additional information such as symbology, percent transparency, queries that select specific features, and joins or relates that link attributes in external tables to the spatial data. You can create layers in ArcCatalog and place them in a shared folder on the network where everyone can access them. Others can add those layers to maps without having to know how to access the database, how to classify data, or even the format in which the data is stored.

Similarly, you can customize metadata in the Catalog to include information that's specific to your organization such as whether or not new coverages have passed quality assurance tests. You can also customize how metadata is displayed so the right information is presented to the right people in your organization.

Overall, ArcCatalog will revolutionize how you do your work. It makes accessing and managing geographic data so easy that before long it will be a constant companion on your desktop.

Tips on learning ArcCatalog

This book introduces ArcCatalog and its capabilities. The topics covered assume you are familiar with the fundamentals of geographic information systems (GIS). If you're new to GIS or feel you need to refresh your knowledge, please take some time to read *Getting Started with ArcGIS,* which you received in your software package. It may not be necessary to read it in its entirety before continuing, but you should use it as a reference if you encounter tasks with which you are unfamiliar.

You don't have to know everything about the Catalog to get immediate results. To begin learning ArcCatalog, read Chapter 2, 'Quick-start tutorial'; there you'll see how easy it is to locate and manage your maps and data. ArcCatalog comes with the data used in the tutorial, so you can follow along at your computer. You can also read the tutorial without using your computer.

This book is designed so that when you have a question, you can get the answer quickly and then complete your task. Although you can read this book from start to finish, you'll likely use it more as a reference. When you need to know how to do a particular task, such as defining a shapefile's coordinate system, look it up in the table of contents or index. You'll find a concise, step-by-step description of how to complete the task. Some chapters also include detailed information that you can read if you want to learn more about the concepts behind the tasks. Refer to the glossary if you come across unfamiliar GIS terms.

Getting help on your computer

In addition to this book, the ArcCatalog online Help system is a valuable resource for learning how to use the software. To learn how to use Help on your computer, see 'Getting help' in Chapter 3 of this book.

Contacting ESRI

If you need to contact ESRI for technical support, see the product registration and support card you received with ArcCatalog or refer to 'Contacting Technical Support' in the 'Getting more help' section of the ArcGIS Desktop Help system. You can also visit ESRI on the Web at *www.esri.com* and *www.arconline.esri.com* for more information on ArcCatalog and ArcInfo™.

ESRI education solutions

ESRI provides educational opportunities related to geographic information science, GIS applications, and technology. You can choose among instructor-led courses, Web-based courses, and self-study workbooks to find education solutions that fit your learning style and pocketbook. For more information, go to *www.esri.com/education*.

Quick-start tutorial

2

ArcCatalog lets you explore and manage your data. After connecting to your data, use the Catalog to explore its contents. When you find the data you want to use, you can add it to a map. Often when you get data for a project, you can't use it right away; you may need to change its projection or format, modify its attributes, or link geographic features to attributes stored in another table. When the data is finally ready, you should document its contents and the changes you have made. These data management tasks can all be accomplished using tools that are available in the Catalog.

The easiest way to find out what you can do with ArcCatalog is to complete the exercises in this tutorial.

- Exercise 1 shows you how to build your own catalog of geographic data by adding data to and removing data from the Catalog.

- Exercise 2 illustrates how to explore and search for data and how to add it to a map.

- Exercises 3 and 4 show you how to define a data source's coordinate system, modify its contents, join attributes in another table to the data, and symbolize features based on the joined attributes. Exercise 3 is for ArcView® and ArcEditor™ users, while Exercise 4 is targeted for ArcInfo users.

This tutorial is designed to let you work at your own pace. You'll need between two and four hours of focused time to complete the tutorial. However, you can also perform the exercises one at a time if you wish.

Exercise 1: Building a catalog of geographic data

When you build a catalog, you're choosing the data you want to work with. You might use several folders of data to complete your project, while someone else might use data stored in a geodatabase. For this tutorial, you'll be working with data from Yellowstone National Park, which is located in the northwestern part of the United States.

In this exercise, you'll add the folder containing the tutorial data to the Catalog. Because you modify this data in later exercises, you'll create a working copy of the folder and then remove items that you don't need from the Catalog.

Start ArcCatalog

Before you can complete the tasks in this tutorial, you must start ArcCatalog.

1. Click the Start button on the Windows taskbar.
2. Point to Programs.
3. Point to ArcGIS.
4. Click ArcCatalog. The ArcCatalog window appears.

What's in the Catalog?

On the left of the ArcCatalog window, you see the *Catalog tree*; it gives you a bird's-eye view of how your data is organized. On the right are tabs that let you explore the contents of the selected item in the Catalog tree.

The first time you start ArcCatalog, it contains *folder connections* that let you access your computer's hard disks. The Catalog also contains folders that let you create and store connections to databases and Internet servers and manage geocoding services and search results.

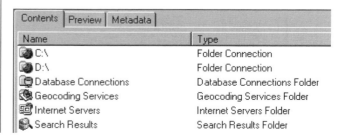

When you select a connection, you can access the data to which it's linked. Folder connections let you access *folders*, or directories, on local disks or shared folders on the network. Database connections let you access the contents of a database. When you remove folder or database connections from the Catalog, you are only removing the connection, not deleting the data.

Together, your connections create a catalog of geographic *data sources*. Individual folders and data sources are *items* in that catalog. If you use ArcInfo Workstation, you're accustomed to using the term "item" when referring to a coverage's attributes; in this book, "item" refers only to an element in the Catalog tree.

Look in a folder connection

When you select a folder connection in the Catalog tree, the Contents tab lists the items it contains. Unlike Windows Explorer, the Catalog doesn't list all files stored on disk; a folder might appear empty even though it contains several files. Folders containing geographic data sources have a different icon to make your data easier to find.

Look in a folder connection in your catalog.

1. Click a folder connection in the Catalog tree. The items it contains appear in the Contents tab. Here, the work folder doesn't contain geographic data; it has a plain folder icon. The icon used by the Workspace folder shows that it contains geographic data.

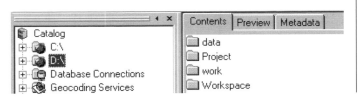

2. Double-click a folder in the Contents list. That folder is selected in the Catalog tree, and the Contents tab lists the folders and geographic data it contains.

Using this method, you can browse through the contents of disks looking for geographic data.

Locate the tutorial folder

Before you can start exploring the data for this tutorial, you must select the folder where the tutorial data is located. In the Catalog, you can quickly select any folder on your computer or on the network as long as you know its path.

1. Click in the Location text box.

2. Type the path to the ArcGIS\ArcTutor folder on the local drive where you installed the tutorial data; for example, type "C:\ArcGIS\ArcTutor".

   If the data was installed by your system administrator in a shared folder on the network, the path to the tutorial folder includes the names of the computer and the share through which the folder is accessed; for example, "\\dataserver\public\ArcGIS\ArcTutor".

3. Press Enter. The folder is selected in the Catalog tree.

When the Catalog already contains a folder connection that can access the ArcGIS\ArcTutor folder, that connection expands and the tutorial folder is selected in the Catalog tree. Otherwise, a new folder connection is added to the Catalog that directly accesses the ArcGIS\ArcTutor folder.

The path you typed above is added to the Location list after you press the Enter key. To access the ArcGIS\ArcTutor folder again in the future, you can choose its path from the list by clicking the Location dropdown arrow.

Create a working copy of the tutorial data

In Exercises 2, 3, and 4, you will create new items and modify the data provided for this tutorial. When processing data, it is best to work on a copy so that your original data will remain unmodified. To prepare for those exercises, use ArcCatalog to copy the ArcGIS\ArcTutor\Catalog folder to a location where you have write access. You will need 15 MB of free disk space to store the tutorial data.

1. If the Catalog doesn't have a connection to the place where you want to store the tutorial data, type its name into the Location combo box and press Enter; for example, type "C:\". A new folder connection will be added to the Catalog. Substitute the name of your folder connection for "C:\" in the following steps.

2. Click the ArcGIS\ArcTutor folder or folder connection in the Catalog tree.

3. Click the Catalog folder in the Contents tab.

4. Click the Copy button.

Copy Paste

5. Click the C:\ folder connection in the Catalog tree.

6. Click the Paste button. A new folder called Catalog will appear in the Contents list for the C:\ folder connection.

7. Click the new Catalog folder in the Contents tab.

8. Click the File menu and click Rename.

9. Type "Cat_Tutorial" and press Enter.

   If you type another name for the folder, that name must have 13 characters or less and cannot use spaces. Because the tutorial data includes coverages, the folder's name must satisfy these requirements for ArcInfo workspaces. For the rest of the tutorial, substitute the name of your folder for "C:\Cat_Tutorial".

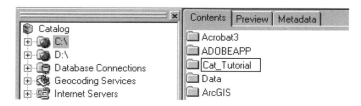

Now that you have a working copy of the tutorial data, you can connect directly to it in the Catalog.

Connect directly to your copy of the tutorial data

Folder connections in the Catalog can access specific folders on disk. You might establish several connections to different folders on the same disk. You don't have to see all the data on your C:\ drive, for example, just because you want to use data in two of the folders it contains.

To create a connection that directly accesses a folder, you can use a shortcut provided by the Catalog.

1. Click the Cat_Tutorial folder in the Contents list.

2. Scroll to the top of the Catalog tree.

3. Drag the Cat_Tutorial folder from the Contents tab and drop it on the Catalog at the top of the Catalog tree.

A new folder connection is added to the Catalog.

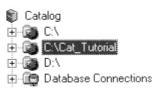

Creating folder connections using this shortcut is handy when you're browsing through local disks that contain many folders, some of which contain geographic data.

Remove folders that you don't need

For the remaining tasks in this tutorial, you only need to use the folder connection that accesses your working copy of the tutorial data. You can remove all other folder connections from the Catalog. To hide folders such as Database Connections, you must change the settings in the Catalog's Options dialog box.

1. Click the C:\ folder connection.

2. Click the Disconnect From Folder button. The connection is removed from the Catalog.

Disconnect From Folder

3. Repeat steps 1 and 2, removing each folder connection in turn, except for the C:\Cat_Tutorial folder connection.

4. Click the Tools menu and click Options.

5. Click the General tab.

6. In the top level entries list, uncheck Database Connections, Internet Servers, and Geocoding Services. Coordinate Systems is unchecked by default.

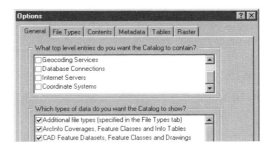

7. Click OK.

Only the C:\Cat_Tutorial folder connection and the Search Results folder remain in the Catalog. Now you can explore the tutorial data without seeing folders and data that aren't essential to the task at hand.

You can continue on to Exercise 2 or stop and complete the tutorial at a later time. If you do not move on to Exercise 2 now, do not delete your working copy of the tutorial data and do not remove the folder connection that accesses the working copy of the tutorial data from the Catalog.

Exercise 2: Exploring data and adding it to a map

Each of the Catalog's three tabs provides a different way to explore the contents of the selected item in the Catalog tree. Within each tab there are different views that let you change the appearance of the selected item's contents.

The Contents tab lists the items contained by the selected item in the Catalog tree, for example, the items in a folder. When a data source such as a shapefile is selected, the Preview tab lets you see the geographic or attribute data it contains. The Metadata tab lets you see documentation describing the item's contents.

ArcCatalog and ArcMap work together to make it easy to build maps. For your project you are mapping the forest resources in the southeastern corner of Yellowstone National Park. In the Yellowstone folder is a map of the study area; it is incomplete. In this exercise, after exploring the data in the folder, you'll add more data to the map.

The Contents tab

When you select items such as folders or geodatabases in the Catalog tree, the Contents tab lists the items they contain. To change the appearance of the Contents list, use the appropriate buttons on the Standard toolbar.

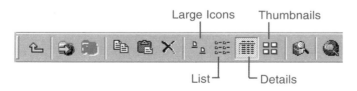

Large Icons view represents each item in the list with a large icon. List view uses small icons. Details view shows

properties of each item in columns; you can sort the list by the property values. Thumbnails view displays a snapshot for each item in the list, providing a quick illustration of the item's geographic data.

Items such as maps, shapefiles, and tables don't contain other items. When you select them in the tree, the Contents tab lists the item's properties and its thumbnail.

Explore the contents of the Yellowstone folder

Each type of geographic data has its own set of icons in the Catalog. The Yellowstone folder contains a personal geodatabase, coverages, shapefiles, raster datasets, a TIN dataset, a dBASE® table, and an ArcMap map document. Geodatabases let you store spatial data inside a relational database; personal geodatabases can be accessed by only one person at a time. For more information, see Chapter 3, 'Catalog basics'.

The Yellowstone folder also contains two layers. A *layer* includes a shortcut to geographic data and information such as the symbology used to draw geographic data on a map, the query used to select which features the layer represents, and properties defining how those features are labeled.

Use the Contents tab to find out about the data in the Yellowstone folder.

1. Double-click the C:\Cat_Tutorial folder connection in the Catalog tree.

2. Click the Yellowstone folder in the Catalog tree.

3. Click the Large Icons button.

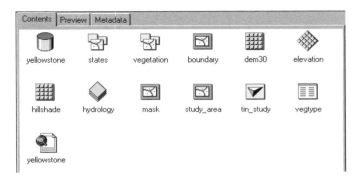

4. Click the List button.

5. Click the Details button. The Type column in Details view will help you learn which icon represents which type of data. You can find more information about data types and their icons in Chapter 4, 'What's in the Catalog'.

6. Click the heading of the Type column to sort the items by type.

7. Click the heading of the Name column to sort the items by name.

Items that contain other items, such as geodatabases and coverages, always appear at the top of the Contents list. They are grouped by type.

8. Click the Thumbnails button and scroll down through the snapshots. Items that are not geographic data sources, such as the yellowstone geodatabase, can't have thumbnails; their icons appear on a gray background.

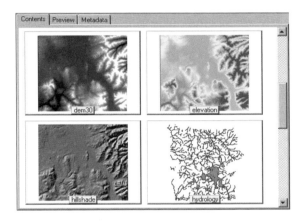

9. Click the Details button.

10. Double-click the yellowstone personal geodatabase in the Contents tab. It contains a feature dataset called water and a feature class called roads.

11. Double-click the water feature dataset in the Contents tab. It contains three feature classes called lakes, rivers, and streams, respectively.

12. Click the states coverage in the Catalog tree to list the feature classes it contains.

13. Click the tin_study TIN dataset in the Catalog tree. This surface represents the terrain of the study area. Because a TIN dataset doesn't contain other items, the Contents tab lists its properties and thumbnail instead.

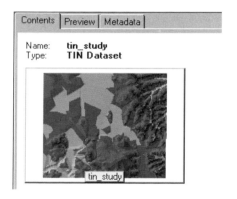

Thumbnails give you a quick look at an item's geographic data; they're useful when browsing through folders. However, you must often see the data in more detail to determine whether or not you want to use it.

The Preview tab

The Preview tab lets you explore the selected item's data in either Geography or Table view. For items containing both geographic data and tabular attributes, you can toggle between the Geography and Table views using the dropdown list at the bottom of the Preview tab.

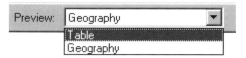

Geography view draws each feature in a vector dataset, each cell in a raster dataset, and each triangle in a TIN dataset. When drawing geographic data, the Catalog uses a default set of symbology. When drawing a layer's contents, the Catalog uses the symbology stored in the layer. You can explore the selected item's geographic data using the buttons on the Geography toolbar.

Table view draws all rows and columns and the value for each cell in the selected item's table. You can explore the values in the table using the scroll bars, the buttons at the bottom of the table, and the context menus that are available from the column headings.

Look at the Yellowstone data in Geography view

Use Geography view to look at the data contained by the items in the Yellowstone folder. When using Geography view, the Geography toolbar is enabled. You can use the Zoom In, Zoom Out, Pan, Full Extent, and Identify buttons on the toolbar to explore geographic data.

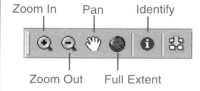

While you look at the Yellowstone data, you may wish to see the geographic features in more detail using these tools.

For more information about how these buttons work, see Chapter 7, 'Exploring an item's geography'.

1. Click the dem30 raster dataset in the Catalog tree. This dataset contains elevation information for the study area within the park.

2. Click the Preview tab. The raster draws using the default grayscale color ramp.

3. Click the elevation layer in the Catalog tree. The raster draws using the symbology stored in the layer—a green to red color ramp.

4. Click the study_area shapefile in the Catalog tree. It represents the study area for this project, which lies in the southeastern corner of the park.

5. In the Catalog tree, click the plus sign next to the vegetation coverage. The coverage's feature classes are listed in the Catalog tree.

6. In the Catalog tree, click each feature class in the vegetation coverage in turn and look at their contents in Geography view. This polygon coverage represents the different types of vegetation within the study area.

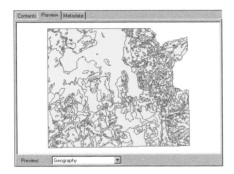

7. Click the Preview dropdown arrow at the bottom of the Preview tab and click Table.

8. In the Catalog tree, click each feature class in the vegetation coverage in turn and look at their contents in Table view.

   All feature classes in a coverage have FID and Shape columns. They may also have several *pseudo items* whose names begin with a dollar sign ($), which are maintained by ArcInfo. Because topology only exists for the polygon feature class, it's the only one that has a feature attribute table and therefore additional attributes.

9. Click the Preview dropdown arrow at the bottom of the Preview tab and click Geography.

10. In the Catalog tree, click the plus sign next to the water feature dataset in the yellowstone geodatabase. The feature classes it contains appear in the Catalog tree.

11. In the Catalog tree, click each feature class in the water feature dataset in turn and look at their contents in Geography view. The feature dataset groups feature classes that contain different types of water features throughout the park: lakes, rivers, and streams.

12. Click the lakes feature class in the Catalog tree.

13. Click the Identify button on the Geography toolbar and click one of the larger lakes in the Preview tab. Its attributes appear in the Identify Results window. Only the larger lakes in this feature class have names.

All features within three pixels of where you clicked are selected; each feature is listed on the left of the Identify Results window. If a data source has no text attributes, a numeric attribute will be used to identify the features.

14. Click the Close button to close the Identify Results window.

15. Click the hydrology layer in the Catalog tree. This group layer presents all features in the lakes, rivers, and streams feature classes using symbology stored in the layer. When added to a map, a group layer only has one entry in the table of contents.

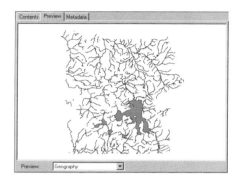

By exploring a data source's data in Geography view, you can find out if it has the features you need and if those features have the correct attributes. This information can help you decide whether or not to add the data to a map.

Explore the contents of a table

With the table exploration tools available in the Catalog, you can learn a great deal about a table's contents. You can search for values in the table and sort records according to the values in one or more columns.

The vegetation coverage represents areas with distinct types of vegetation. The vegtype table contains both general information, such as whether or not the area is forested, and detailed information such as which plant species live in each area. Throughout the park, 67 vegetation groups have been defined. Some areas, such as open water or rocky peaks, may not have any vegetation at all.

Use the tools available in Table view to explore the contents of the vegtype table.

1. Click the vegtype dBASE table in the Catalog tree.

2. Explore the table's values using the buttons at the bottom of the table. Once you click in the table you can also use the arrow keys on the keyboard.

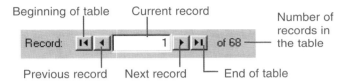

Beginning of table   Current record   Number of records in the table
Record: [◄◄] [◄]   1   [►] [►►] of 68   End of table
Previous record   Next record

3. Type "10" into the Current record text box and press Enter. The Current record icon appears next to the tenth row in the table. The Object ID value for this record, which is located in the OID column, is nine; the OID values begin at zero.

4. Scroll horizontally through the table until you see the column named "Primary".

5. Right-click the heading of the Primary column and click Freeze/Unfreeze Column. The Primary column is frozen in position at the left of the table, and a heavy black line appears to its right.

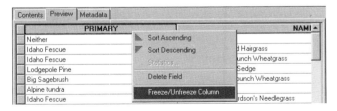

6. Scroll horizontally through the table. The Primary column stays in place while the other columns scroll normally. Place the Type column to the right of the Primary column.

7. Position the mouse pointer over the right edge of the Primary column's heading. The mouse pointer changes.

8. Click and drag the edge of the Primary column's heading to the left. The red line indicates the edge's original position, while the black line shows its new location. Drop the edge of the column. The column is narrower.

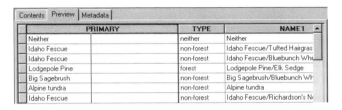

9. Right-click the heading of the Type column and click Freeze/Unfreeze Column. The heavy black line is now to the right of the Type column. Scroll horizontally through the table. Both the Primary and Type columns stay in place while the other columns scroll normally.

10. Click the Type column's heading. The column is highlighted in light blue.

11. Click and drag the Type column's heading to the left of the Primary column. The red line indicates the Type column's new position. Drop the column in the new location.

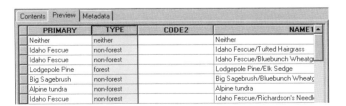

The Type column now appears to the left of the Primary column, and the heavy black line appears to the right of the Primary column.

12. Hold down the Ctrl key and click the Primary column's heading. Both columns are now selected.

13. Right-click the heading of the Primary column and click Sort Descending.

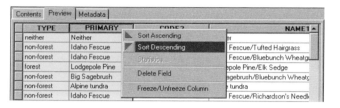

The rows in the table are sorted alphabetically in descending order, first by the values in the Type column, then by the values in the Primary column. In this format, the table presents vegetation information from general to detailed as you look at the columns from left to right.

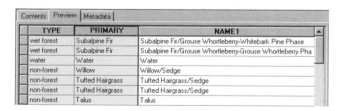

After exploring the contents of the vegtype table, you have a better idea about the forest resources that are available in the study area. However, you may not know what some column names in the table mean. For that information, you must look at the table's metadata.

The Metadata tab

The Metadata tab shows descriptive information about the selected item in the Catalog tree. Metadata includes properties and documentation. Properties are derived from the data source, while documentation is information provided by a person.

Metadata is stored as *extensible markup language (XML)* data in a file with the data or in a geodatabase. The Catalog uses an *extensible stylesheet language (XSL)* stylesheet to transform the XML data into a *hypertext markup language (HTML)* page. You can change the metadata's appearance by changing the current stylesheet using the dropdown list on the Metadata toolbar.

You can browse through the available metadata just as you would browse through any Web page in a browser.

Explore metadata for the tutorial data

The metadata for the Yellowstone folder provides an overview of the data it contains. By looking at metadata for the study_area shapefile and vegetation coverage, you can find out when and why the data was created and look at the data's properties.

1. Click the Yellowstone folder in the Catalog tree.

2. Click the Metadata tab. A metadata HTML page describing the contents of the Yellowstone folder appears.

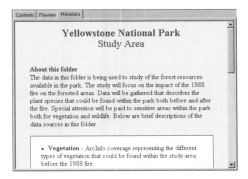

3. Click the study_area shapefile in the Catalog tree. Metadata is presented with the ESRI stylesheet by default.

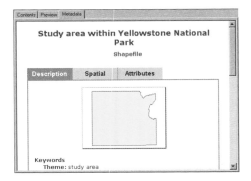

The Catalog automatically adds the item's current property values to the metadata when you view it in the Metadata tab.

4. Click the Stylesheet dropdown arrow on the Metadata toolbar and click FGDC. Click the Stylesheet dropdown arrow again and click FGDC FAQ. Different stylesheets present the same body of metadata in different ways.

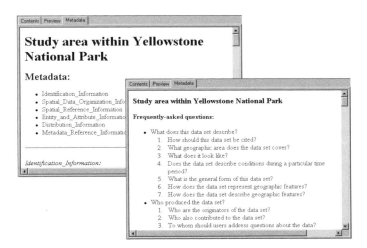

5. Click the Stylesheet dropdown arrow on the Metadata toolbar and click ESRI.

6. Click the Spatial tab in the metadata.

7. Underneath the name of the coordinate system used by the data, click Details. The coordinate system's properties appear. Click Details again to hide the information.

8. Scroll down to see the shapefile's extent. Its bounding coordinates are presented both in the actual coordinates of the data and in decimal degrees.

9. Click the boundary shapefile in the Catalog tree.

10. Click the Spatial tab in the metadata. No coordinate system information is available because the shapefile's projection hasn't yet been defined. Scroll down to see the data's extent in projected coordinates.

Because there is no coordinate system information, the Catalog cannot calculate the extent in decimal degrees. If you have ArcView or ArcEditor, you'll define the shapefile's coordinate system in Exercise 3.

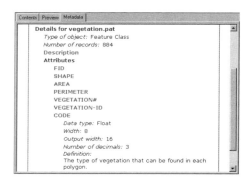

By default, when you look at metadata in the Metadata tab, the Catalog automatically creates metadata if it doesn't already exist. On creation, it adds the current property values along with hints about the type of documentation you should provide. These hints appear as gray text when using the ESRI stylesheet.

While metadata isn't required to meet any standard, the metadata created by the Catalog will follow the Federal Geographic Data Committee (FGDC) *Content Standard for Digital Geospatial Metadata*. If all the suggested documentation is completed, your metadata will meet the FGDC standard's minimum requirements.

11. Click the vegetation coverage in the Catalog tree.

12. Click the Spatial tab in the coverage's metadata. Like the boundary shapefile, this coverage's projection hasn't been defined, and the data's extent in decimal degrees hasn't been calculated. If you have ArcInfo, you'll define its coordinate system in Exercise 4.

13. Click the Attributes tab in the metadata. Attributes are listed for each coverage feature class that has them.

14. Scroll down to and click the CODE attribute. You can see a description of its data type and values.

15. Click the Contents tab in the ArcCatalog window.

Now that you've familiarized yourself with the data, you can create a map describing the study area within Yellowstone National Park.

Add a layer to a map

The Yellowstone folder contains a map document. The map already has data from the roads feature class and the hillshade raster dataset. You must still add the water features in the park to the map.

Adding data to a map is easy—all you have to do is drag data from the Catalog and drop it on the map. When you drop a layer onto a map, a copy of the layer is created and stored inside the map document. In this way, you can create a layer once and use it in many different maps. Before you can add more data to the yellowstone map, you must open the map document in ArcMap.

1. Click the Yellowstone folder in the Catalog tree.

2. Double-click the yellowstone map document in the Catalog; it opens in ArcMap.

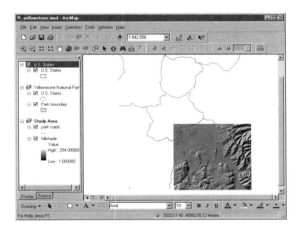

In the ArcMap window, you see the contents of the Study Area data frame in Data view; the name of the active data frame appears in bold in the table of contents. The map has two other data frames as well: Yellowstone National Park and United States.

3. Arrange the ArcMap and ArcCatalog windows so you can see the table of contents in ArcMap and the Catalog tree at the same time.

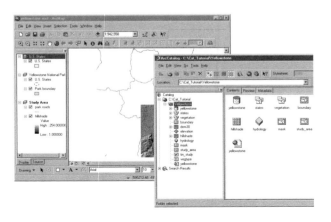

Water features must be added to the map; they should draw underneath the park roads layer but above the hillshade layer. The map's table of contents reflects the order in which layers are drawn.

4. Click and drag the hydrology layer from the Catalog and drop it in the map's table of contents below park roads in the Study Area data frame's list of layers.

5. In the ArcMap window, click the Save button.

The water features draw underneath the road features and above the hillshade image in the map.

Create layers

The features in the park roads and hydrology layers cover the entire park, but you only want to map the study area. The mask shapefile represents the area outside the study area. You can create a layer from this shapefile in the Catalog and add it to the map to hide features that extend beyond the study area.

If you add data directly to a map rather than creating a layer first, ArcMap creates a new layer in the map document. After modifying the layer's symbology and other properties, you can save it as a layer file outside the map document so you can reuse it in other maps.

In this task, you will first save the park roads layer to a file, then create a new layer representing the mask shapefile.

1. Right-click the park roads layer in the map's table of contents and click Save As Layer File.

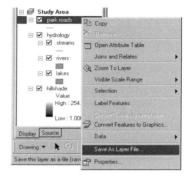

2. In the Save Layer dialog box, navigate to the Yellowstone folder. Type a name for the new layer, for example, "park roads", then click Save.

3. In the Catalog tree, click the Yellowstone folder.

4. Click the View menu and click Refresh. The park roads layer now appears in the Contents list.

5. Right-click the mask shapefile and click Create Layer.

6. Navigate to the Yellowstone folder in the Save Layer As dialog box. Type a name for the new layer, for example, "feature mask", then click Save. The layer appears in the Contents list for the Yellowstone folder.

7. Right-click the feature mask layer and click Properties. The Layer Properties dialog box appears.

8. Click the Symbology tab.

9. Click the button showing the current symbol for the layer. A symbol is assigned at random when the layer is created.

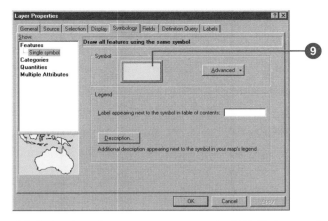

10. Click the dropdown arrow next to the Fill Color symbol and click white.

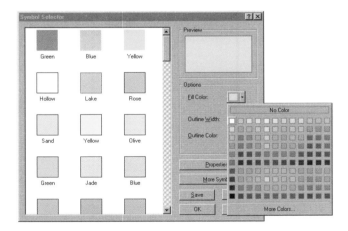

11. Click the dropdown arrow next to the Outline Color symbol and click black.

12. Click OK in the Symbol Selector dialog box and click OK in the Layer Properties dialog box.

Now that you've finished creating the feature mask layer, you should document its contents. Before adding it to the map, you can import metadata, which has already been created for the layer.

Import metadata

There is a text file in the Yellowstone folder containing metadata describing the feature mask layer. By importing this metadata it will become part of the layer itself; it will be copied and moved with the data and updated automatically by ArcCatalog. This is useful because when metadata is maintained separately, it is easy to let it get outdated.

1. Click the feature mask layer in the Catalog tree.

2. Click the Metadata tab. By default, ArcCatalog automatically creates metadata for the feature mask layer. Information such as the name of the layer file will be included automatically along with documentation hints.

3. Click the Import metadata button on the Metadata toolbar.

Import metadata

4. Click the Format dropdown list and click FGDC CSDGM (TXT). The metadata in the text file is structured following the format, which is supported by the FGDC's metadata parser, mp.

5. Click Browse and navigate to the Yellowstone folder.

6. Click the Files of type dropdown arrow and click All files (*.*). Click the file "feature_mask.met" and click Open.

7. Click OK in the Import Metadata dialog box.

The documentation in the feature_mask.met file has been added to the layer's metadata, replacing the Catalog's documentation hints.

Now there is just one thing missing from the layer's metadata—a thumbnail describing how the layer will appear when added to a map. Creating thumbnails for data sources and layers is a manual process.

8. Click the Contents tab. Instead of a thumbnail, the layer's icon appears on a gray background.

9. Click the Preview tab.

10. Click the Create Thumbnail button on the Geography toolbar.

Create Thumbnail

11. Click the Contents tab. You can see the thumbnail both here and in the layer's metadata.

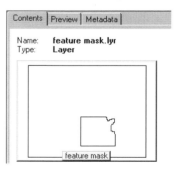

After adding or removing features from a data source or changing a layer's symbology, you may wish to update the item's thumbnail.

Search for items

You've explored the contents of the Yellowstone folder and created the new feature mask layer, and soon you'll add it to your map. However, often you know what data you need to use but not where it's located. The Catalog lets you search for data by its name, type, and geographic location. You can also search for data using dates and keywords that reside in the data's metadata. When metadata exists, its name, type, and geographic location are derived from the metadata as well.

Suppose you didn't know the feature mask layer existed. You need to add a data source to the yellowstone map that shows the boundary of the study area within the park.

1. Right-click the Yellowstone folder in the Catalog tree and click Search. The Search dialog box appears, and the location in which the search will begin looking is automatically set to the Yellowstone folder.

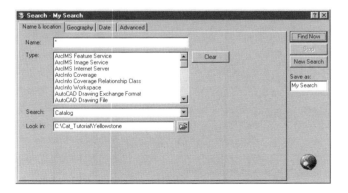

2. Click the Advanced tab.

3. Click the Metadata element dropdown arrow, scroll down, and click Theme Keyword in the list.

4. Click the Condition dropdown arrow and click equals.

5. Click in the Value text box and type "boundary".

6. Click Add to List.

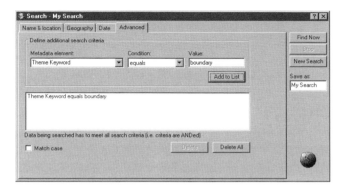

With this search, ArcCatalog will look in the Yellowstone folder for items whose metadata includes the theme keyword "boundary".

7. Click Find Now.

The search is saved in the Search Results folder and is selected in the Catalog tree. As the items are found that satisfy the search criteria, shortcuts to those items are added to the search's list of results. When a Search is complete, the message "Catalog search finished" appears in the status bar of the ArcCatalog window and the Stop button becomes unavailable in the Search dialog box.

8. Click the Close button in the upper-right corner of the Search dialog box.

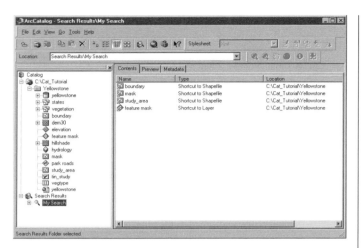

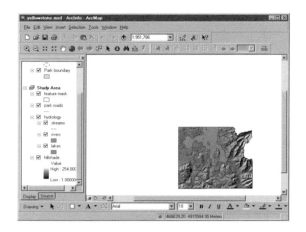

The Catalog has found four items in the Yellowstone folder that satisfy the search criteria: three shapefiles and the feature mask layer. Shortcuts to these items appear in the Contents tab. In the Catalog, you can work with shortcuts the same way you would work with the items themselves.

9. Click the shortcut to the feature mask layer in the Contents tab.

10. Click the Preview tab to draw the layer.

11. Click the Metadata tab to see the layer's metadata.

   This is the data you want to add to your map.

12. Click and drag the shortcut to the feature mask layer from the Catalog and drop it in the map's table of contents above park roads in the Study Area data frame's list of layers. Now you only see features inside the study area on the map.

13. Click the Save button.

The yellowstone map now contains all the basic data needed to represent the study area. Now all you need to do is add a few finishing touches to the map's layout.

Complete the map

The purpose of the yellowstone map is to illustrate the different types of vegetation within the study area. Currently, the Study Area data frame shows most of the park. You need to zoom in on the study area; do this in layout view to make sure you're zoomed in as far as possible but still able to see the entire study area in the layout. When you switch to layout view, you'll see the Yellowstone National Park and United States data frames as well.

1. Click the View menu and click Layout view. You can see all the data frames and the map's title and scale bar in the layout.

In the Yellowstone National Park data frame, there is an orange rectangle representing the area that you can see within the Study Area data frame. Similarly, in the United States data frame there is a green rectangle representing the area that you can see within the Yellowstone National Park data frame.

When you zoom farther in to the features in the Study Area data frame, the extent rectangle on the Yellowstone National Park data frame decreases in size and the scale bar increases in size.

2. In ArcMap, click the Zoom In button on the Tools toolbar and draw a rectangle around the study area in the Study Area data frame.

Zoom In   Go Back To Previous Extent

If you zoom in too close, you can return to your previous extent by clicking the Go Back To Previous Extent button.

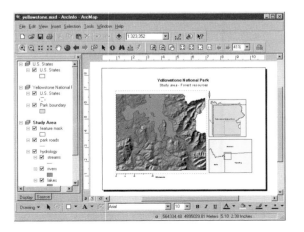

3. When the study area fills the Study Area data frame, click the Save button.

4. Click the File menu and click Exit to stop ArcMap.

5. In ArcCatalog, click the Contents tab and click the Yellowstone folder in the Catalog tree.

In this exercise, you saw how to explore your data in the Catalog, create layers, import metadata, search for items, and add them to maps. In the next exercise, you'll create a layer illustrating the different types of vegetation that can be found within the study area. This requires changing a data source to a different format and modifying its values.

If ArcView or ArcEditor is installed on your computer, you should continue on to Exercise 3; you'll define a shapefile's coordinate system, modify its contents, and then use it to create the vegetation type layer. If ArcInfo is installed on your computer, you should continue with Exercise 4; you'll define a coverage's coordinate system, modify its contents, and then use it to create the vegetation type layer. The title bar of the ArcCatalog window shows you which version of the ArcGIS™ software you have installed on your computer.

You can continue on to the next exercise or stop and complete the tutorial at a later time. If you do not move on to the next exercise now, do not delete your working copy of the tutorial data. Also, do not remove the folder connection that accesses the working copy of the tutorial data from the Catalog.

Exercise 3: Managing shapefiles

Assembling data for a project often requires significant data management work. In this exercise, ArcView and ArcEditor users will create a layer representing the different types of vegetation in the study area and add it to the yellowstone map. In doing so, you'll learn how to define a shapefile's coordinate system, modify attributes, join a table's attributes to a shapefile, and update their metadata using tools that are available in the Catalog. If ArcInfo is installed on your computer, skip to Exercise 4.

Define a shapefile's coordinate system

In the previous exercise, when you looked at metadata for the boundary shapefile you found that its coordinate system was not defined. The features in the shapefile are projected, but the Catalog doesn't know which map projection was used. Without that information, the Catalog can't determine where on the earth's surface the features are located.

A shapefile's Properties dialog box lets you modify its attributes, create spatial and attribute indexes, and define its projection.

1. In the Contents list, right-click the boundary shapefile and click Properties.

2. Click the Fields tab.

3. Under Field Name, click Shape. This column contains the feature geometry. The shapefile's spatial properties appear in the Field Properties list below.

    At the bottom of the list is the Spatial Reference property. The shapefile's coordinate system is unknown.

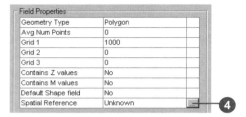

4. Click the ellipses (...) button to the right of the Spatial Reference property.

    All data sources in the Yellowstone folder except for the states coverage use the same projection. You can copy coordinate system information from any data source in the folder except states to this shapefile.

5. Click Import in the Spatial Reference Properties dialog box.

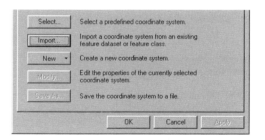

6. Navigate to the Yellowstone folder, click the dem30 raster dataset, then click Add. The projection parameters from the dem30 raster dataset appear in the Spatial Reference Properties dialog box.

7. Click OK. The shapefile's coordinate system appears in the Shape column's Spatial Reference property.

8. Click OK.

   A shapefile's coordinate system information is stored in a .prj file with the data—for example, boundary.prj. Now you can update the shapefile's metadata with the new coordinate system information. By default, every time you look at metadata in the Metadata tab, the Catalog automatically updates the metadata with the data source's current properties.

9. Click the Metadata tab.

10. Click the Spatial tab in the metadata.

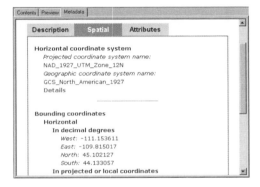

The Catalog has updated the coordinate system in the metadata and has calculated its extent in decimal degrees.

Modify attributes in dBASE tables

A layer can join or relate attributes in a table of any format to its geographic data source as long as they share a column of values. The only requirement is that the columns have the same data type. You must look at the vegetation coverage's attributes and the vegtype dBASE table's

columns to see if modifications are required before they can be joined together.

Using Table view, you can see that the vegtype dBASE table has two columns that contain no values and that the attribute named "CODE" in the vegetation coverage's polygon feature class and the column named "VEGID" in the vegtype table contain the same values. Use the coverage's metadata to see the data type of the CODE attribute, then use the table's Properties dialog box to modify its columns appropriately.

1. Click the vegetation coverage in the Catalog tree.

2. Click the Attributes tab in the metadata.

3. Scroll down to and click the CODE attribute. Its data type is Float.

4. Click the Contents tab in the ArcCatalog window.

5. Right-click the vegtype table in the Catalog tree and click Properties.

6. Click the Fields tab. The table's columns and their data types are listed. You can see that the VEGID column's data type is Long Integer.

   The values in the VEGID column in the dBASE table are integers, while the CODE column in the coverage contains floating point values. To join the table to the coverage, the values in both columns must have the same data type. Therefore, you must add a floating point column to the table.

7. Scroll to the bottom of the list of column names. Click in the empty row under the name of the last attribute and type "VEGTYPE".

8. Click under Data Type to the right of the new column's name, click the dropdown arrow that appears, then click Float.

9. In the Field Properties list below, click to the right of Precision, click again, then replace the zero with "4".

10. Click to the right of Scale and replace the zero with "1".

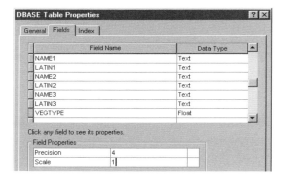

The new attribute has been defined. Now you can remove the empty columns from the table.

11. Point at the gray button to the left of the NAME3 column; the mouse pointer changes to an arrow. Click to select the column.

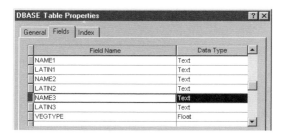

12. Press Delete. The column is removed from the list.

13. Repeat steps 11 and 12 for the LATIN3 column to delete it from the table.

14. Click OK.

Now that the vegetation shapefile has a new integer attribute, you must copy the values in the original VEGID attribute to the new VEGTYPE attribute. To edit the values in a table, you must use ArcMap.

Calculate attribute values in ArcMap

To edit a table's values you must add the table to a map. With the Editor toolbar visible and the table opened, you can start editing its values. Use the field calculator to copy values from the CODE attribute to the VALUE attribute.

1. Click the Launch ArcMap button in ArcCatalog.

Launch ArcMap

2. Click OK to start using ArcMap with a new, empty map.

3. Drag the vegtype table from the Catalog and drop it onto the table of contents or the canvas in the ArcMap window.

4. Click the Source tab in the map's table of contents. The table's data is available within the map.

5. Click the Editor Toolbar button; the Editor toolbar appears.

Editor Toolbar

6. On the Editor toolbar, click the Editor menu and click Start Editing.

7. Right-click the vegtype table and click Open. The table's values appear in a table window. The headings of the columns whose values you can change have a white background.

8. Right-click the heading of the VEGID column and click Freeze/Unfreeze Column. It is now positioned at the left of the table with a heavy black line to its right.

9. Scroll horizontally in the table until you see the VEGTYPE column.

10. Right-click the heading of the VEGTYPE column and click Freeze/Unfreeze Column. It is now positioned to the right of the VEGID column at the left of the table with the heavy black line to its right.

11. Right-click the heading of the VEGTYPE column and click Calculate Values.

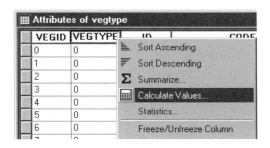

12. Scroll down and double-click VEGID in the Fields list. "[VEGID]" appears in the text box below "VEGTYPE =". ArcMap reads this as VEGTYPE=[VEGID]; all values in the VEGTYPE column will be set equal to the values stored in the VEGID column.

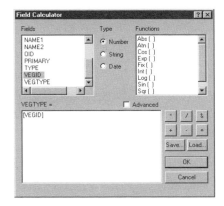

13. Click OK.

As ArcMap calculates the new values in each record, you can see how many records have been completed at the bottom of the Field Calculation dialog box.

14. Look at the VEGTYPE column in the table; its contents are the same as the VEGID column.

15. Click the Close button in the top-right corner of the table window.

16. Click the Editor menu on the Editor toolbar and click Stop Editing. Click Yes to save your changes.

17. Click the File menu and click Exit to stop ArcMap. Click No when prompted to save the map.

Now that the vegetation coverage and the vegtype table have matching columns, you can create a layer that links the two together.

Create a layer using the related attributes

Through a layer you can join the attributes stored in the vegtype table to the vegetation coverage and use the table's values to query, label, and symbolize the coverage's features.

1. Click the Yellowstone folder in the Catalog tree.

2. Click the File menu, point to New, then click Layer.

3. Type a name for the layer such as "vegetation type".

4. Click the Browse button, navigate to the Yellowstone folder, click the vegetation coverage, then click Add.

5. Check Store relative path name, then click OK. Relative path names let you continue using the layer even after moving or renaming the Yellowstone folder.

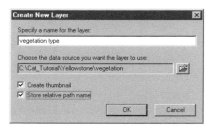

6. Right-click the vegetation type layer and click Properties.

7. Click the Joins & Relates tab in the Layer Properties dialog box.

8. Click the Add button next to the Joins list.

9. Click the first dropdown arrow to specify what you want to join to this layer, then click Join attributes from a table.

10. Under step 1, click the dropdown arrow and click the CODE attribute.

11. Under step 2, click the Browse button.

12. Navigate to the Yellowstone folder, click the vegtype table, and click Add.

13. Under step 3, click the dropdown arrow and click the VEGTYPE column.

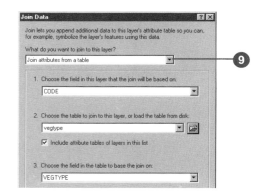

14. Click OK in the Join Data dialog box. The vegtype table is added to the list of tables that have been joined to the coverage.

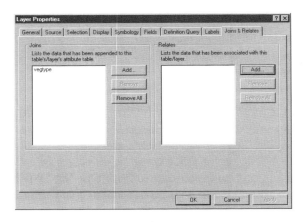

15. Click the Symbology tab.

16. Click Categories in the Show list.

17. Click the Value Field dropdown arrow and click vegtype.TYPE.

18. Click the Color Scheme dropdown arrow and click a different color palette, if desired.

19. Click Add All Values.

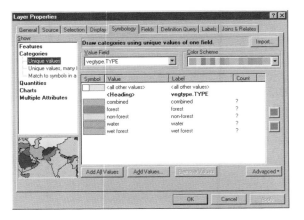

To change the color of individual values, for example, to make the water polygons blue, double-click the patch of color to the left of the value. Set the fill and outline colors in the Symbol Selector dialog box and click OK.

20. Click OK in the Layer Properties dialog box.

The vegetation type layer represents the forest resources in the study area.

Add the vegetation type layer to the map

Now that the vegetation layer has been created, you can add it to the yellowstone map.

1. Double-click the yellowstone map in the Catalog.

2. Click and drag the vegetation type layer from the Catalog and drop it in the map's table of contents below park roads and above hydrology in the Study Area data frame's list of layers.

   A message appears indicating that the vegetation type layer is missing spatial reference information. ArcInfo users define the vegetation coverage's coordinate system in Exercise 4.

3. Click OK to add the vegetation type layer to the map. Because the vegetation coverage's features share the same projection as the other data sources in the Study Area data frame, the vegetation features draw on the map in the proper location even though their coordinate system hasn't yet been defined.

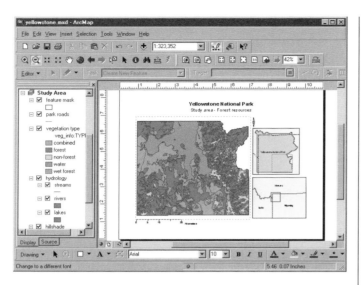

4. Click the Save button in ArcMap.

5. Click the File menu and click Exit to stop ArcMap.

The yellowstone map is now complete.

You're now finished with the Quick-start tutorial. This exercise showed how to use ArcCatalog to manage your shapefiles and dBASE tables. You saw how to define a shapefile's coordinate system, add and remove attributes, calculate attribute values, and symbolize features using attributes stored in another table.

Overall, this tutorial has introduced you to a wide range of tasks. Whether you are looking for data, building maps, or managing data for your project or an entire organization, the Catalog plays a pivotal role in getting the job done. The remaining chapters in this book look in detail at the variety of tasks you can accomplish with the Catalog.

Exercise 4: Managing coverages

Assembling data for a project often requires significant data management work. In this exercise, ArcInfo users will create a layer representing the different types of vegetation in the study area and add it to the yellowstone map. In doing so, you'll learn how to define a coverage's coordinate system, add and delete attributes, update metadata, and link a coverage and an associated INFO table using tools that are available in the Catalog.

In addition to ArcInfo Desktop, you must have ArcInfo Workstation or a geoprocessing server installed on your computer to be able to complete this exercise. ArcView and ArcEditor users must follow Exercise 3 instead.

Set a coverage's coordinate system

In Exercise 2, when you looked at metadata for the vegetation coverage you found that its coordinate system was not defined. The features in the coverage are projected, but the Catalog doesn't know which map projection was used. Without that information, the Catalog can't determine where on the earth's surface the features are located.

In addition to creating topology and setting tics, extents, and tolerances for a coverage, you can define its coordinate system using the Coverage Properties dialog box. You can copy coordinate system information to and from coverages, *grid*-based raster datasets, and TIN datasets. The dem30 and the tin_study data sources in the Yellowstone folder use the same projection as the vegetation coverage.

1. Click the Yellowstone folder in the Catalog tree.

2. In the Contents list, right-click the vegetation coverage and click Properties.

3. Click the Projection tab and click Define.

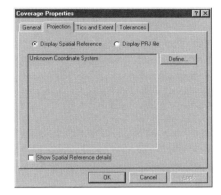

4. Click Define a coordinate system for my data to match existing data and click Next.

5. Click the Browse button to the right of the Dataset text box.

6. Navigate to the Yellowstone folder, click the dem30 raster dataset, then click Open.

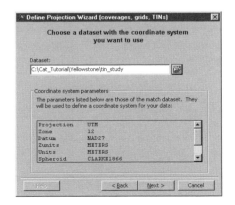

The projection parameters used by the raster dataset appear in the Coordinate system parameters list.

7. Click Next and click Finish.

8. The name of the projection and the *spheroid* or *ellipsoid* it uses appears in the Properties dialog box. To list the projection's parameters, check Show spatial reference details. Click Display PRJ file to see the parameters in ArcInfo Workstation format.

9. Click OK.

Now that the coverage's projection has been defined, you won't have any trouble adding its data to any map, regardless of the map's projection.

Convert your data to a different format

ArcToolbox provides many different tools for converting your data. The most common data conversion tools have been included in the Catalog. If the conversion tool you need isn't listed, start ArcToolbox and look in the Conversion Tools toolset.

1. In the Contents list, right-click the vegtype dBASE table.

2. Point to Export and click dBASE to INFO. The vegtype table is automatically listed as the Input dBASE table.

3. Click the Browse button to the right of the Output INFO table text box.

4. Navigate to the Yellowstone folder. Type a name for the new table—for example, "veg_info"—into the Name text box and click Save.

5. Click OK. A dialog box appears showing the progress of the conversion tool. When the tool has finished running, the dialog box disappears and the new veg_info INFO table appears in the Yellowstone folder.

Now you must look at the vegetation coverage's attributes and the veg_info table's columns to see if other modifications are required before they can be linked together.

Modify attributes in coverages and INFO tables

The relationship between the vegetation coverage and the veg_info table is established by creating a coverage relationship class. A relationship class is similar to an ArcInfo *relate,* but it lets you define the relationship more accurately. Once created, a relationship class lets you query, label, and symbolize features in the coverage using attributes in the associated table.

When creating the relationship class, you define which column in the feature class's attribute table and which column in the INFO table share the same values. The columns must also share the same data type.

Using Table view, you could see that the vegetation coverage's polygon feature class has a column named "CODE", and the veg_info table has a column named "VEGID" that contains the same values. Use the Properties dialog box for the feature class and table to see if the columns have the same data type.

1. Right-click the veg_info INFO table in the Catalog and click Properties.

2. Click the Items tab. The columns in the table and their data types are listed. The VEGID column has a data type of Binary and an input width of four.

    Because this column will be used to join the table's values to the coverage, it should be indexed.

3. Click the VEGID attribute and click Add Index. The value in the Indexed column in the Properties dialog box changes from No to Yes.

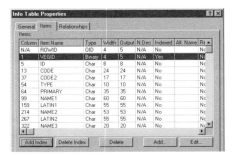

If you looked at the contents of the original vegtype or the new veg_info table, you would see that the NAME3

and LATIN3 columns don't contain any values and can be deleted.

4. Scroll down to and click the NAME3 attribute, then click Delete. Click the LATIN3 attribute and click Delete. The columns are removed from the list.

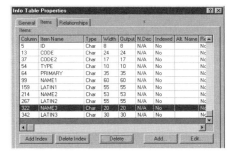

5. Click OK.

6. Click the vegetation coverage in the Catalog tree.

7. Right-click the polygon feature class in the Contents list and click Properties. The Properties dialog box for a coverage feature class is the same as for INFO tables.

8. Click the Items tab. The attributes in the feature class and their data types are listed. The CODE attribute has a data type of Float.

    The values in the VEGID column in the INFO table are binary integers, while the CODE column in the polygon feature class contains floating point values. To create a relationship class the values in both columns must have the same data type. Therefore, you must add a Binary column to the polygon feature class.

9. Click Add.

10. Click next to Item name and type "VEGTYPE".

ArcInfo

11. Click next to Type, click the dropdown arrow that appears, then click Binary. The Input width defaults to four, and the Display width defaults to five.

12. Click OK. The new attribute appears in the list of attributes for the feature class.

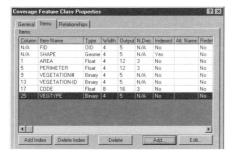

13. Click OK.

Now that the vegetation coverage has a new integer attribute, you must copy the values in the original "CODE" attribute to the new "VEGTYPE" attribute. To edit the features and attributes in a coverage, you must use ArcMap.

Calculate attribute values in ArcMap

You must add a coverage feature to a map in order to edit its attributes. With the Editor toolbar visible, you can start editing the coverage. Open the feature class's attribute table and then use the field calculator to copy values from the CODE attribute to the VEGTYPE attribute.

1. Click the Launch ArcMap button in ArcCatalog.

Launch ArcMap

2. Click OK to start using ArcMap with a new, empty map.

3. Drag the vegetation coverage from the Catalog and drop it onto the table of contents or the canvas in the ArcMap window. A layer is created for the coverage's polygon feature class and added to the map's table of contents.

4. Click the Editor Toolbar button; the Editor toolbar appears.

Editor Toolbar

5. On the Editor toolbar, click the Editor menu and click Start Editing.

6. Right-click the vegetation polygon layer in the table of contents and click Open Attribute Table. The attributes of the polygon feature class appear in a table window. The headings of the columns whose values you can change have a white background.

7. Scroll horizontally in the table until you see the VEGTYPE column.

8. Right-click the heading of the VEGTYPE column and click Calculate Values.

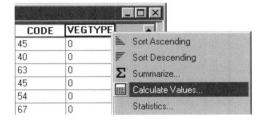

9. Double-click CODE in the Fields list. "[CODE]" appears in the text box below "VEGTYPE =". ArcMap reads this as VEGTYPE=[CODE]; in other words, all values in the VEGTYPE column will be set equal to the values stored in the CODE column. Click OK.

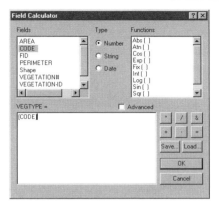

As ArcMap calculates the new values in each record, you can see how many records have been completed at the bottom of the Field Calculator dialog box.

10. Look at the VEGTYPE column in the attribute table; its contents are the same as the CODE column.

11. Click the Close button in the top-right corner of the attribute table window.

12. On the Editor toolbar, click the Editor menu and click Stop Editing. Click Yes to save your changes.

13. Click the File menu and click Exit to stop ArcMap. Click No when prompted to save the map.

Now that the vegetation coverage and the veg_info table have matching columns, you can create a relationship class that links the two together.

Create a relationship class

When creating a relationship class, you must define the properties of the relationship between the coverage and the associated INFO table. One property of a relationship is its *cardinality*, which describes how many features in the

ArcInfo

coverage are related to how many records in the attribute table. The cardinality for this relationship is many-to-one; that is, a group of plants defined in the veg_info table may be found in many polygons in the coverage.

Because you're primarily interested in the vegetation coverage's features, the polygon feature class is the *origin* in the relationship. The veg_info table is the *destination*. The columns that link the two are called key attributes. The VEGTYPE attribute in the coverage is the *primary key*; the VEGID column in the veg_info table is the *foreign key*.

Path labels describe the relationship when navigated forward and backward. In this relationship, going forward from the origin to the destination, the vegetation polygons relate to "types of vegetation" in the veg_info table. Going backward from the destination to the origin, the veg_info table relates to "vegetation features" in the coverage.

Now, create a relationship class linking the vegetation coverage's polygon feature class to the veg_info table.

1. Click the Yellowstone folder in the Catalog tree.

2. Click the File menu, point to New, then click Coverage Relationship Class.

3. Type a name for the relationship class, for example, "vegetation_attributes".

4. In the Origin table/feature class list, click the plus sign next to vegetation and click its polygon feature class.

5. In the Destination table/feature class list, click veg_info.

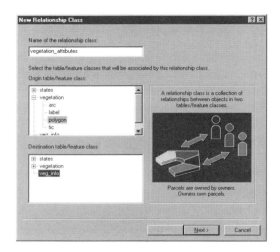

6. Click Next.

7. Click Simple (peer-to-peer) relationship and click Next.

8. In the first text box, type "types of vegetation". In the second text box, type "vegetation features". Click Next.

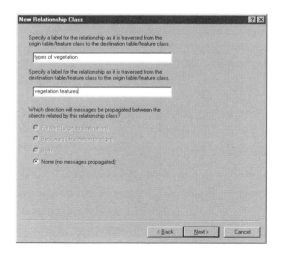

As discussed at the beginning of this task, there is a many-to-one relationship between the polygons in the vegetation coverage and the vegetation types in the veg_info table. In practice, a many-to-one relationship is the same as a one-to-one relationship. Therefore, you will create a relationship with one-to-one cardinality.

9. Click 1-1 (one-to-one) and click Next.

10. Click the dropdown arrow for the primary key field list and click the VEGTYPE attribute.

11. Click the dropdown arrow for the foreign key field list and click VEGID.

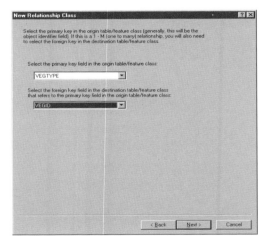

12. Click Next and click Finish. The vegetation_attributes coverage relationship class appears in the Yellowstone folder.

You can find out which relationships a coverage feature class or INFO table participates in by opening its Properties dialog box and clicking the Relationships tab. Now you can create a layer that takes advantage of this relationship.

Create a layer using the related attributes

Once a relationship class has been created, a layer can use the attributes in the veg_info table to query, label, and symbolize the vegetation coverage's features.

1. Click the Yellowstone folder in the Catalog tree.

2. Click the File menu, point to New, then click Layer.

3. Type a name for the layer such as "vegetation type".

4. Click the Browse button, navigate to the Yellowstone folder, click the vegetation coverage, then click Add.

5. Check Store relative path name, then click OK. Relative pathnames let you continue using the layer even after moving or renaming the Yellowstone folder.

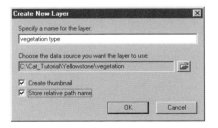

6. Right-click the vegetation type layer and click Properties.

ArcInfo

7. Click the Joins & Relates tab in the Layer Properties dialog box.

8. Click the Add button next to the Joins list.

9. Click the first dropdown arrow to specify what you want to join to this layer, then click Join data based on a predefined relationship class.

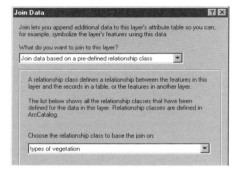

10. Click OK in the Join Data dialog box. The veg_info table is added to the list of tables that have been joined to the coverage.

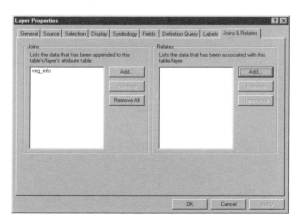

11. Click the Symbology tab.

12. Click Categories in the Show list.

13. Click the Value field dropdown arrow and click veg_info:TYPE.

14. Click the Color Scheme dropdown arrow and click a different color palette, if desired.

15. Click Add All Values.

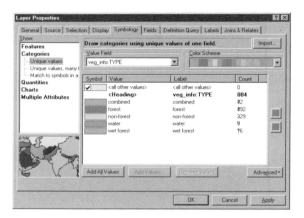

To change the color of individual values—for example, to make the water polygons blue—double-click the patch of color to the left of the value. Set the fill and outline colors in the Symbol Selector dialog box and click OK.

16. Click OK in the Layer Properties dialog box.

The vegetation type layer joins attributes in the veg_info table to the polygons in the vegetation coverage using information from the vegetation_attributes relationship class. Then, it symbolizes them using the table's values. It represents the forest resources, which are available throughout the study area.

Add the vegetation type layer to the map

Now that the vegetation layer has been created, you can add it to the yellowstone map.

1. Double-click the yellowstone map in the Catalog.
2. Click and drag the vegetation type layer from the Catalog and drop it in the map's table of contents below park roads and above hydrology in the Study Area data frame's list of layers.

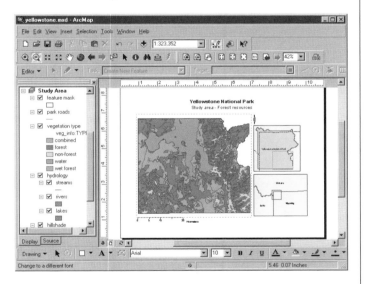

3. Click the Save button in ArcMap.
4. Click the File menu and click Exit to stop ArcMap.

The yellowstone map is now complete.

Update the coverage's metadata

Over the course of this exercise, you defined the vegetation coverage's coordinate system, added a new attribute, and created a relationship class linking the coverage to the veg_info table. Now you should update the vegetation coverage's metadata with this information.

1. Click the vegetation coverage in the Catalog tree.
2. Click the Metadata tab. The Catalog has automatically added the coverage's coordinate system to the metadata and has calculated its extent in decimal degrees.
3. Click the Attributes tab in the metadata.
4. Click the VEGTYPE attribute. The Catalog has added this attribute and its data type information to the list.
5. Scroll down to see information about the relationship class in which the coverage participates. Relationship class information isn't included in the FGDC standard; ESRI has defined a profile of that standard that includes additional data properties.

   To add descriptive information, for example, to document the data contained in the VEGTYPE column, use the Catalog's FGDC metadata editor.

6. Click the Edit Metadata button on the Metadata toolbar.

Edit Metadata

7. Click Entity Attribute at the top of the metadata editor. The editor is currently showing you metadata for the polygon feature class.

8. Within the Detailed Description tab, click the Attribute tab.

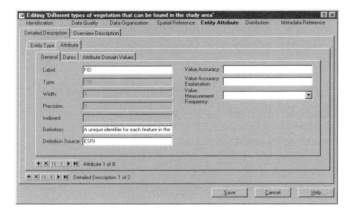

The Attribute tab currently shows metadata for the first attribute in the polygon feature class, the FID attribute. The Definition text box shows that this column contains a unique value for each feature in the coverage.

9. In the toolbar at the bottom of the Attribute tab, click the Move next button. The current attribute displayed in the Attribute tab advances to the next one listed in the coverage's metadata—the SHAPE attribute, which contains feature geometry.

Move next   Move last

10. Click the Move last button to see the VEGTYPE attribute.

11. Click in the Definition text box and type, "A number identifying the type of vegetation in each feature. Use this attribute to join this coverage to the veg_info table."

12. Click Save.

13. Click the Attributes tab in the metadata.

14. Click the VEGTYPE attribute. The description has been added to the metadata.

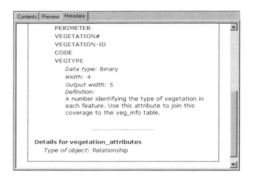

Anyone who uses this coverage in the future will be able to find out what its properties are, what data it contains, and to which tables it is related.

This exercise showed how you can use ArcCatalog to manage your data. You saw how to define a coverage's coordinate system, add and remove attributes, calculate attribute values, symbolize features using attributes in a related table, and use the FGDC metadata editor to document your data.

Overall, this tutorial has introduced you to a wide range of tasks. Whether you are looking for data, building maps, or managing data for your project or an entire organization, the Catalog plays a pivotal role in getting the job done. The remaining chapters in this book look in detail at the variety of tasks you can accomplish with the Catalog.

Catalog basics

3

Whether you need to find a specific map, document its contents, or modify a coverage, you can do it in ArcCatalog. Before you begin, read this chapter. It runs through the basics of what you see when you start the Catalog and how to use it to browse through your data holdings. It also explains how to use the online Help system, which can help you learn you what the various elements are in the ArcCatalog window. Help topics also provide step-by-step instructions that show you how to accomplish your tasks.

Starting ArcCatalog

Starting ArcCatalog is the first step to exploring your data. However, before you can begin, ArcCatalog must be installed on your computer. If you don't know whether it has been installed, check with your system administrator or install it yourself using the installation guide. Once the software is installed, you can access ArcCatalog from the Start button on the Windows taskbar.

Tip

Starting ArcCatalog from ArcMap

You can also start ArcCatalog from ArcMap by clicking the Launch ArcCatalog button on the Standard toolbar.

1. Click the Start button on the Windows taskbar.

2. Point to Programs.

3. Point to ArcGIS.

4. Click ArcCatalog.

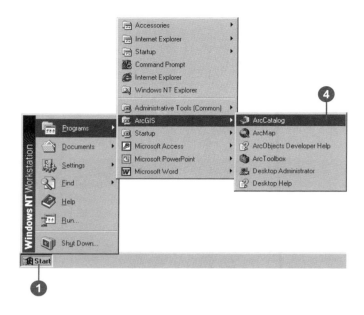

The ArcCatalog window

The title bar displays the selected item's location.

Click the buttons on the active toolbars to explore the current view.

The Catalog tree lets you access all of the Catalog's contents. Select an item in the tree to view its contents.

Each tab displays the contents of the selected item in a different way. Within each tab, several different views let you change how you see the selected item's contents.

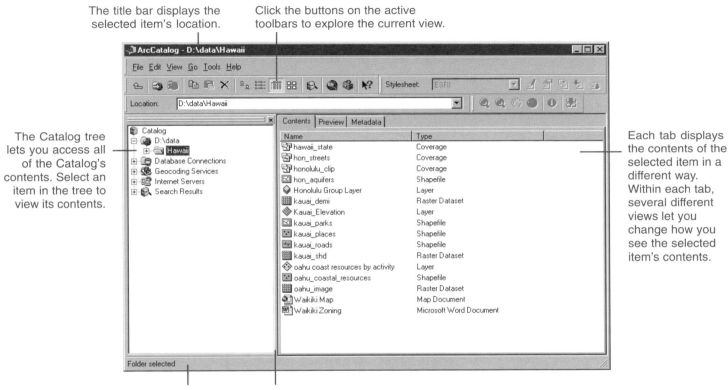

The status bar describes what a button or menu command does and also reports on which items are selected.

Move this bar to resize the Catalog tree.

Browsing through the Catalog

Browsing through geographic data in the Catalog is easy, whether you're looking for something specific or simply seeing what's there. The Catalog tree displays all the items in the Catalog and illustrates the hierarchy in which they are organized. When you select an item in the Catalog tree, the Contents tab lists the items it contains. By pressing the Up and Down arrow keys on your keyboard, you can quickly select the items displayed above or below the currently selected item in the Catalog tree. The Up One Level button on the Standard toolbar selects the next item up in the Catalog tree's hierarchy.

Sometimes, the Catalog's list of a folder's contents will not match the folder's actual contents—for example, after you receive a map as an e-mail attachment and you save it in a folder. You can update, or refresh, the Catalog's list of the folder's contents by selecting the folder in the Catalog tree, clicking the View menu, then clicking Refresh.

When you know exactly which item you want to work with and where it's located, it's quicker to select it by typing its path than by browsing through the Catalog's ▶

Selecting items in the Catalog tree and the Contents list

1. Click the Contents tab.

2. Click an item in the Catalog tree that contains a subset of items (such as a folder or feature dataset).

   The items it contains are listed in the Contents tab.

3. Double-click an item in the Contents list that contains a subset of items.

   The items it contains are listed in the Contents tab.

4. Double-click an item that contains a subset of items in the Catalog tree.

   The items it contains are listed in the Catalog tree.

5. Double-click an item that does not contain other items (such as a shapefile or table) either in the Contents list or in the Catalog tree.

   Maps open in ArcMap, and XML documents and file types open in the appropriate application. A Properties dialog box will appear for all other items.

6. Click an item that doesn't contain other items in the Catalog tree.

   Its properties and thumbnail are listed in the Contents tab.

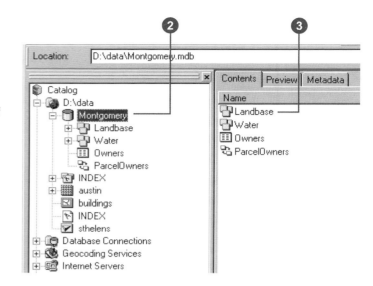

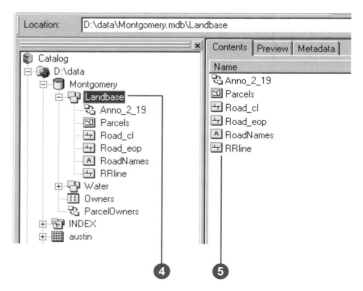

contents. An item's *path* describes how to navigate the Catalog's hierarchy to find it. For items located in a remote database stored in the Database Connections folder, the path includes both the folder's name and the name of the database connection. For example, a path that selects a parcel's feature class in an ArcSDE™ geodatabase might appear as "Database Connections\myConnection\ theDataset\anOwner.parcels". If you type the path of a folder for which a folder connection doesn't already exist, a new folder connection will automatically be created.

The selected item's path always appears in the ArcCatalog window's title bar and in the text box on the Location toolbar. The status bar shows how many items are currently selected and what type of items they are, if appropriate. ▶

Tip

Selecting a path from the Location list

After selecting an item by typing its path, the path is added to the Location list. To select that item again later, click the Location dropdown arrow and click the appropriate path in the list.

Selecting an item by typing its path

1. Type the path to the item you want to select in the Location text box.

2. Press Enter.

   The Catalog tree expands to show the item you've selected. Or a new folder connection is added to the Catalog and it is selected in the Catalog tree.

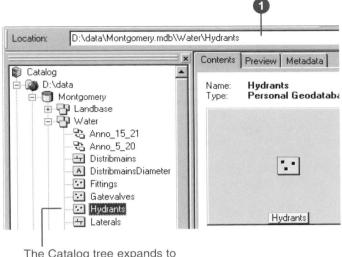

The Catalog tree expands to show the selected item.

Moving up the Catalog tree

1. Click the Up One Level button.

   The next item up in the Catalog tree's hierarchy is selected.

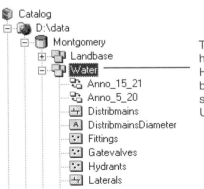

The next item up in the heirarchy is selected; if Hydrants was selected before, Water would be selected after clicking the Up One Level button.

When browsing through a long list of folders, you may find it difficult to locate your data. By default, all folders have the same, plain folder icon. To make your data easier to find, the Catalog gives you the option to use a special icon for folders that directly contain GIS data. With this option checked, you will experience delays when accessing data across the network. After locating the folder you need, add a folder connection that directly accesses it, then uncheck the option to use the special icons. Most illustrations in this book show the Catalog with this option checked.

Tip

Returning to your last location

By default, when ArcCatalog is started it automatically selects the item that was selected the last time you stopped ArcCatalog. This option is convenient, but if you turn it off ArcCatalog may start faster, especially if the item is located on the network.

Tip

Hiding file extensions

By default, file extensions are hidden for maps, layers, file types, and raster datasets. When file extensions are shown, a raster would be listed as "Buffalo.bil", for example.

Making it easier to find geographic data on disk

1. Click the Tools menu and click Options.

2. Click the General tab.

3. Check Use a special icon for folders containing GIS data.

4. Click OK.

   Folders that directly contain GIS data will have a special folder icon to make your data easier to find.

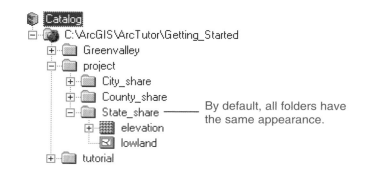

By default, all folders have the same appearance.

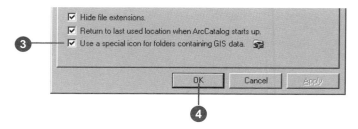

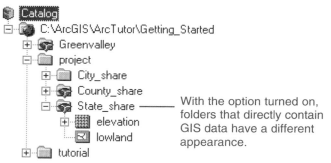

With the option turned on, folders that directly contain GIS data have a different appearance.

Repositioning the Catalog tree

The main elements of the ArcCatalog window are the Catalog tree, which displays the contents of the Catalog, and the tabs, which provide different views of the contents of the selected item in the Catalog tree. You can reposition the Catalog tree to make it easier to explore the selected item's contents. By default, the Catalog tree is docked on the left side of the ArcCatalog window, but you can dock it on the right, the top, or the bottom of the window if you prefer. You can also undock the Catalog tree so that it floats on your desktop to make it easier to drag datasets to a map or a Geoprocessing tool. Similarly, you might want to hide or close the Catalog tree temporarily.

Undocking the Catalog tree

1. Click and drag the bar at the top of the Catalog tree with your mouse pointer to a location outside the ArcCatalog window.

   A rectangle with a thick line indicates where the Catalog tree will be placed.

2. Drop the panel.

   The Catalog tree is floating on the desktop but is still working with the application.

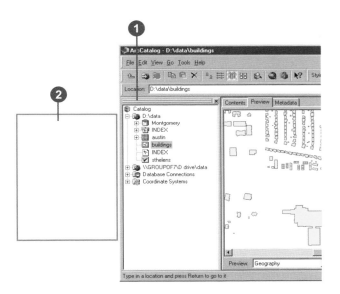

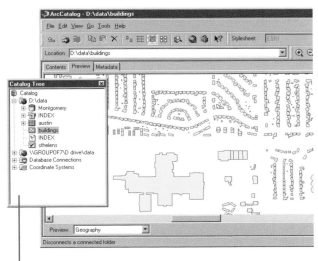

The Catalog tree is floating on the desktop.

Hiding and showing the Catalog tree

1. Click the Close button on the bar at the top of the Catalog tree to hide it.

   The Catalog tree no longer appears in the ArcCatalog window.

2. Click the View menu and check Catalog tree to show it again.

The Catalog tree no longer appears in the ArcCatalog window.

Getting help

A quick way to learn about ArcCatalog is to use the online Help system. You can get help in a variety of ways.

When you position the mouse pointer over a button for a second or two, the button's name pops up in a small box called a ToolTip. Also, when you position the pointer over an element in the ArcCatalog window, a brief description of what it does appears in the status bar. To quickly display a longer description of a button or menu command, click the What's This? button, then click the element in which you're interested. You can also get help in some dialog boxes. When you click the What's This? button in the upper-right corner and click an element in the dialog box, a description of the item pops up. Some dialog boxes have a Help button; clicking it opens a Help topic with detailed information about the task you're trying to accomplish.

Much of the information in this book is available in the online Help system. The Help viewer contains a navigation pane—with Contents, Index, and Search tabs—and a topic pane for viewing Help topics. Both panes are visible in the viewer at all ▶

Getting help in the ArcCatalog window

1. Click the What's This? button on the Standard toolbar.

2. Click the Help pointer on the element in the ArcCatalog window you want to learn more about.

3. Click anywhere on the screen to close the Help description box.

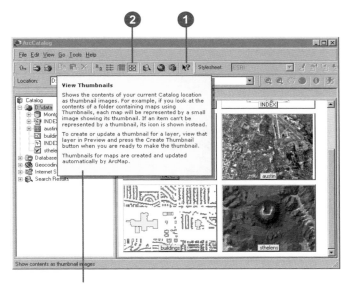

The Help description box shows information about the button.

Getting help in a dialog box

1. Click the What's This? button on the upper-right corner of the dialog box.

2. Click the Help pointer on the element in the dialog box you want to learn more about.

3. Click anywhere on the screen to close the Help description box.

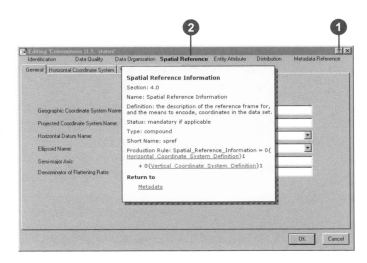

times, allowing you to keep track of where you are in the Help system. The Help topics are organized around the tasks you want to complete; they also provide the concepts behind the tasks. Use the Help Contents to look up general topics and see how they are organized. You can also search the index by entering keywords that identify your task. Alternatively, use the Search tab to find Help topics that have specific words or phrases.

To get more precise results, you can perform an advanced search using wildcard expressions, Boolean operators, or nested expressions. *Wildcard expressions* let you search for one or more characters using a question mark (?) or an asterisk (*). *Boolean operators* include AND, OR, NOT, and NEAR. Use these operators to define which words should be included and excluded in the topics you want to find. *Nested expressions* let you create complex searches for information. For examples of advanced search expressions, see the online Help topic 'Using this Help system'; you'll find it in the 'Getting more help' book in the Contents tab.

Using the Contents tab to get help

1. Click the Help menu and click ArcGIS Help.

2. Click the Contents tab.

3. Click the plus sign next to a book to see a list of topics in it.

4. Click the topic you want to read.

5. View the topic in the topic pane on the right.

6. Click the minus sign next to a book to close it.

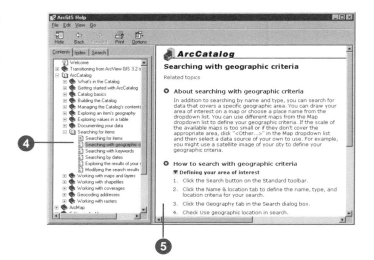

Working with Help topics

When you right-click in the topic pane of the Help viewer you'll see a shortcut menu. It lets you quickly print topics, copy text, or return to a previous Help topic.

Using the Index tab to get help

1. Click the Help menu and click ArcGIS Help.

2. Click the Index tab.

3. In the text box, type one or more keywords representing the topic you want to find.

4. Click the keyword that interests you from the list below, then click Display.

5. If several topics are related to the keyword you selected, the Topics Found dialog box appears. Click the keyword that interests you and click Display.

6. View the topic in the topic pane on the right.

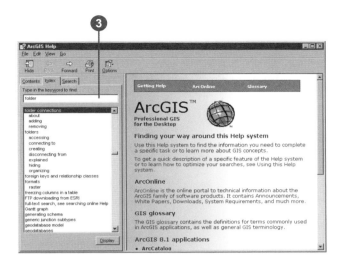

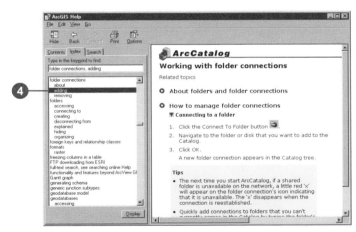

Searching for words within a topic

You can search for words in the topic you're currently viewing. Press Ctrl + F on the keyboard to open the Find dialog box, then type the keyword you want to find. The keyword is highlighted everywhere it appears in the current topic.

Finding Help topics containing specific words

1. Click the Help menu and click ArcGIS Help.

2. Click the Search tab.

3. Type the word or words that should be contained in the topics you want to find.

   Your word expression may include an asterisk (*) to represent several characters, or a question mark (?) to represent one character. It may also use Boolean operators such as AND, OR, NOT, and NEAR.

4. Click List Topics.

5. Double-click the topic you're interested in.

6. View the topic in the topic pane on the right.

   The word or words you specified are highlighted in the topic.

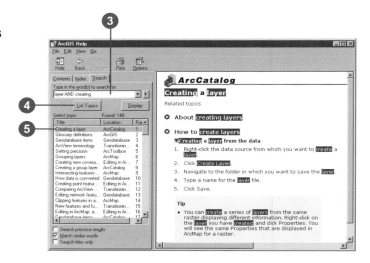

Stopping ArcCatalog

Close ArcCatalog when you're finished working with it. ArcCatalog automatically remembers which folder connections are in the Catalog, which toolbars are visible, and where elements of the ArcCatalog window are positioned. By default, the Catalog also remembers which item was selected in the tree when you closed it and will select that item again the next time you start ArcCatalog.

1. Click the Close button in the ArcCatalog window.

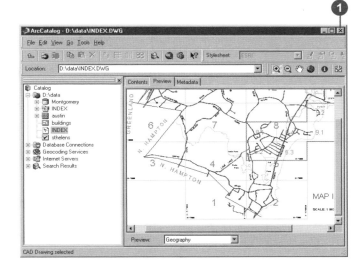

What's in the Catalog?

4

ArcCatalog is the place where you can assemble connections to all the data you need to use. When you select a connection you can access the data to which it's linked. The connection might access a folder on a local disk, a database on the network, or an ArcIMS Internet server. Together, your connections create a catalog of geographic *data sources*.

Individual folders and coverages are *items* in that catalog. If you use ArcInfo Workstation, you're accustomed to using the term "item" when referring to a coverage's attributes; in this book, "item" refers only to an element in the Catalog tree such as a folder.

This chapter briefly describes the items that appear in the Catalog by default. For more information about data formats and how they model features on the earth's surface, read *Modeling Our World*. The data sources supported by ArcCatalog can be extended by programmers to include additional data formats; for information about this, read *Exploring ArcObjects*.

Folders and file types

When you first start ArcCatalog, it contains *folder connections* that let you access your computer's hard disks. You can add folder connections that access specific *folders,* or directories, on a local disk, shared folders on the network, or the contents of a floppy or CD–ROM drive when appropriate. You can also remove folder connections that you don't need.

When you look in a folder connection, you see the items it contains. Unlike Windows Explorer, the Catalog doesn't list all files on disk. A folder might appear empty in the Catalog even though it isn't. Only the items that you choose to see will be included in the folder's contents.

In addition to choosing which geographic data formats you want to work with, you can choose to see files such as Microsoft® Word documents that contain information about your data. To do so, you must add them to the file types list. A *file type* consists of the file extension identifying the file on disk (for example, .doc), the image representing the file type in the Catalog, and a description. See Chapter 5, 'Building the Catalog', for more information.

By default, all folders have the same icon. Instead, you might want to use a different icon to represent folders that contain the items you've chosen to work with in the Catalog. This can make your geographic data easier to find, especially if you have a large number of folders on a local disk. However, with this option you will experience delays when accessing data across the network.

ArcCatalog also has folders that let you manage Database Connections, Geocoding Services, connections to Internet Servers, and Search Results. Another folder for managing Coordinate Systems is hidden by default. These are discussed in greater detail later in the chapter. You can hide these folders if you don't use them.

The next time you start ArcCatalog, a computer on the network, a database, or an Internet server may be unavailable. When you see a little red 'x' on the icon for your folder, database, or Internet server connection, it is disconnected; until your connection is reestablished, their data is inaccessible.

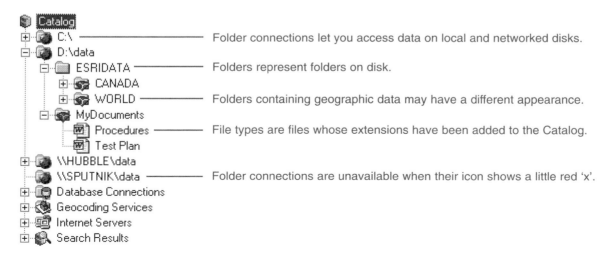

Folder connections let you access data on local and networked disks.

Folders represent folders on disk.

Folders containing geographic data may have a different appearance.

File types are files whose extensions have been added to the Catalog.

Folder connections are unavailable when their icon shows a little red 'x'.

Maps, layers, and graphs

You can access maps, layers, and graphs from ArcCatalog. A *map* is essentially a printed map stored on disk. It can contain geographic data, titles, legends, and north arrows. You can use custom *map templates* to create a series of maps with the same layout. You can also embed map documents in other documents—for example, in Microsoft Word.

Layers include symbology, display, label, query, and relationship information, all of which defines how geographic data is drawn on a map. For example, a layer might select specific cities from a shapefile, draw them as blue squares, and label them with text stored in a related table. Layers don't include the data itself; they reference data sources stored elsewhere.

Layers can be stored either inside a map document or as individual layer files. They are an effective communication tool that can be shared in an organization. For example, you can place predefined layers in a shared folder on the network. By using those layers, others can add data to their maps without having to know where the geographic data is stored, how to join it to a related table, or what each column means.

Several layers can be combined to form a *group layer*. When added to a map, all layers in the group will be represented by one entry in the table of contents. For example, you might create a group layer representing the background material for a map. Similarly, you could combine road, railroad, and ferry shapefiles into a single transportation theme.

In ArcMap, you can create *graphs* illustrating the attributes of features or records in a table. For example, your graph could illustrate population density within the regions that appear on the map. By saving a graph to a file you can add graphs with the same format to a series of maps. When added to a map, the graph establishes a link to the data and will change dynamically to reflect the currently selected features or records.

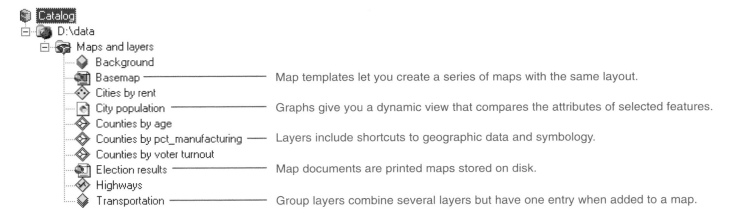

Shapefiles, dBASE tables, and text files

A folder can contain *shapefiles,* which store geographic features and their attributes. Geographic features in a shapefile can be represented with points, lines, or polygons (areas). The folder might also contain dBASE tables, which can store additional attributes that can be joined to a shapefile's features.

All files that have the file extensions .txt, .asc, .csv, or .tab appear in ArcCatalog as *text files* by default. However, in the Options dialog box you can choose which of these file types should be represented as text files and which should not be shown in the Catalog tree. When text files contain comma- and tab-delimited values, you can see those values in Table view and join them to geographic features. Text files are read-only in ArcCatalog.

In a layer's Properties dialog box, the Joins & Relates tab lets you join attributes stored in a dBASE table or text file to the features in a shapefile. If, instead, a table contains information describing spatial locations such as x,y,z coordinates or street addresses, you can create a shapefile representing those locations with tools available in the Catalog.

In addition to shapefiles and tables, ArcView GIS 3 users work with project files, legend files, and Avenue™ scripts. To view these in the Catalog, you must add their file extensions to the Catalog's file types list. For example, to see ArcView GIS 3 projects in the Catalog, add the file extension .apr to the list. Details about adding file types can be found in Chapter 5, 'Building the Catalog'.

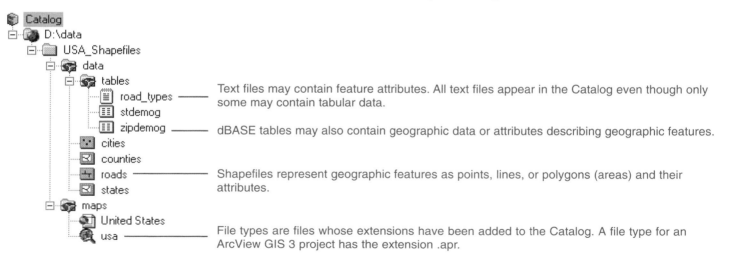

Text files may contain feature attributes. All text files appear in the Catalog even though only some may contain tabular data.

dBASE tables may also contain geographic data or attributes describing geographic features.

Shapefiles represent geographic features as points, lines, or polygons (areas) and their attributes.

File types are files whose extensions have been added to the Catalog. A file type for an ArcView GIS 3 project has the extension .apr.

Coverages and INFO tables

Coverages use a set of *feature classes* to represent geographic features; each feature class stores a set of points, lines (arcs), polygons (areas), or annotation (text). Feature classes can have *topology,* which determines the relationships between features.

More than one feature class is often required to define the features. For example, line and polygon feature classes both exist in a coverage representing polygon features. Polygon features also have label points, which appear as a separate feature class. Every coverage has a feature class containing tic points, which represent known real-world coordinates.

Feature attributes are stored in a separate INFO table for each feature class in the coverage. Other attributes can be stored in INFO or relational database management system (RDBMS) tables, then joined to features with a *relationship class.*

When you look in a folder in the Catalog, you see all the coverages and INFO tables it contains; you don't see the INFO folder itself. Look in a coverage to see its feature classes; each feature class represents both the features and their associated feature attribute table. For example, after selecting a polygon feature class you can preview its features and its attributes.

Coverages often have associated files; to see them in the Catalog, add them as file types. For example, to see ARC Macro Language (AML™) scripts, you would add the file extension .aml to the file types list.

PC ARC/INFO® coverages are like ArcInfo coverages except that their attributes are stored in dBASE tables. PC ARC/INFO coverages can be previewed in the Catalog and you can create metadata for them, but other data management operations such as copy and paste are not available.

Coverages created with ArcInfo before version 7 appear dimmed in the Catalog. After converting the workspace using ArcInfo Workstation, you can access their contents in ArcCatalog.

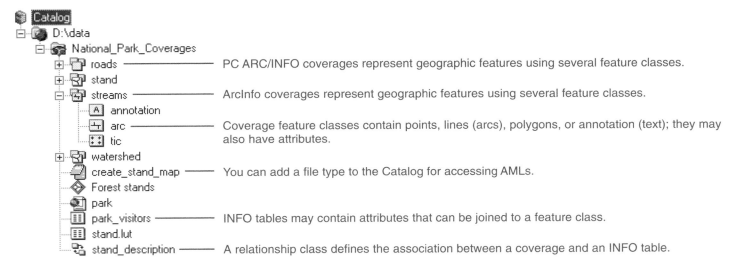

PC ARC/INFO coverages represent geographic features using several feature classes.

ArcInfo coverages represent geographic features using several feature classes.

Coverage feature classes contain points, lines (arcs), polygons, or annotation (text); they may also have attributes.

You can add a file type to the Catalog for accessing AMLs.

INFO tables may contain attributes that can be joined to a feature class.

A relationship class defines the association between a coverage and an INFO table.

Geodatabases

Geodatabases are relational databases that contain geographic information. Geodatabases contain feature classes and tables. Feature classes can be organized into a *feature dataset*; they can also exist independently in the geodatabase.

Feature classes store geographic features represented as points, lines, or polygons and their attributes; they can also store annotation and dimensions. All feature classes in a feature dataset share the same coordinate system. *Tables* may contain additional attributes for a feature class or geographic information such as addresses or x,y,z coordinates.

Many objects in a geodatabase can be related to each other. For example, tables containing customer addresses and billing information are related, just as state and county feature classes are related. To explicitly define the relationships between objects in a geodatabase, you must create a relationship class. Relationships let you use attributes stored in a related object to symbolize, label, or query a feature class.

Feature classes in a feature dataset can be organized into a *geometric network*. The network combines line and point feature classes to model linear networks—for example, electrical networks—and maintains topological relationships between its feature classes.

Creating and accessing geodatabases

To manage your own spatial database, you can create a *personal geodatabase*. If you do this, your data will be stored in a Microsoft Access database. For a multiuser spatial database, use ArcSDE™, which lets many people in an organization simultaneously update data stored in a centrally located RDBMS. SDE® for Coverages lets you access coverage, ArcStorm™, or ArcInfo LIBRARIAN™ databases just as you would an RDBMS.

You can access personal geodatabases directly in the Catalog, but to access data stored in an RDBMS you must add a *database connection*. To do so, double-click one of the Add Database Connection wizards in the Database Connections folder. You'll be prompted for information such as your username and password and the database to which you want to connect.

In general, when you create a database connection, you choose the data provider that will retrieve your data from the database. For multiuser spatial databases, ArcSDE is the data provider.

Object Linking and Embedding Database (OLE DB) providers generally retrieve nonspatial data only such as tables of data. You can preview these tables in the Catalog and join their values to spatial data. If an OLE DB provider can retrieve spatial data and present it in Open GIS Consortium, Inc. (*OpenGIS® or OGC*), format, you can preview that data in the Catalog.

All database connections are stored in the Database Connections folder by default, but they can be moved elsewhere. For example, you can place a connection that has read-only access to the database in a shared folder where others can access it; they can use it to view the database's contents in the Catalog without having to know the details of how to connect to the database.

The first time you select a database connection, ArcCatalog tries to connect to the database. If the connection attempt is successful, ArcCatalog will list the items contained in the database. When you see a little red 'x' on the database connection's icon, it is disconnected; until you have reestablished the connection, you can't access data stored in the database.

Items in an RDBMS are owned by the user who created them. When the Catalog lists the contents of an ArcSDE geodatabase, the owner's name appears before the item's name. For example, a feature class named "valves" that is owned by the user "admin" would appear in the Catalog as "admin.valves".

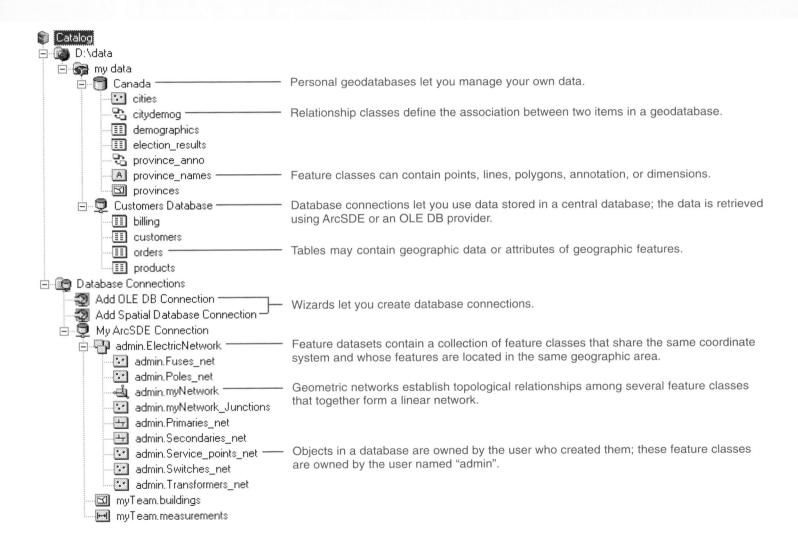

Catalog
D:\data
    my data
        Canada ─────────────── Personal geodatabases let you manage your own data.
            cities
            citydemog ──────────── Relationship classes define the association between two items in a geodatabase.
            demographics
            election_results
            province_anno
            province_names ─────── Feature classes can contain points, lines, polygons, annotation, or dimensions.
            provinces
        Customers Database ─────── Database connections let you use data stored in a central database; the data is retrieved
            billing   using ArcSDE or an OLE DB provider.
            customers
            orders ───────────── Tables may contain geographic data or attributes of geographic features.
            products
Database Connections
    Add OLE DB Connection ──────┐
    Add Spatial Database Connection ─┤── Wizards let you create database connections.
    My ArcSDE Connection
        admin.ElectricNetwork ──────── Feature datasets contain a collection of feature classes that share the same coordinate
            admin.Fuses_net   system and whose features are located in the same geographic area.
            admin.Poles_net
            admin.myNetwork ─────────── Geometric networks establish topological relationships among several feature classes
            admin.myNetwork_Junctions   that together form a linear network.
            admin.Primaries_net
            admin.Secondaries_net
            admin.Service_points_net ── Objects in a database are owned by the user who created them; these feature classes
            admin.Switches_net   are owned by the user named "admin".
            admin.Transformers_net
        myTeam.buildings
        myTeam.measurements

Raster and TIN datasets

ArcCatalog lets you directly access raster data in a variety of formats. You can work with many types of images, ArcInfo grids, and rasters stored in ArcSDE geodatabases. The supported raster file formats are listed in the Catalog's Options dialog box. All rasters are treated the same in ArcCatalog regardless of their format. However, you can make a raster's format apparent by unchecking Hide file extensions in the Options dialog box.

Raster datasets are comprised of one or more raster bands. A *raster band* is a rectangular matrix of cells. Individual grids, such as *digital elevation models (DEM)*, and single-band images appear in the Catalog as raster datasets with one band. Multispectral images appear as raster datasets containing several raster bands. In Table view you can see a table listing a raster band's attributes. The table may include descriptive information about the values or how they are displayed.

Large raster mosaics often have to be divided into tiles that are stored in separate files on disk. When stored in an ArcSDE geodatabase, the individual tiles can be combined to create one seamless raster mosaic. In the Catalog, an ArcSDE raster dataset's name is presented as the owner's name, followed by the raster's name. For example, a raster dataset named "Waterloo_ortho", which is owned by the user "Ryan", would appear in the Catalog as "Ryan.Waterloo_ortho".

Raster catalogs appear in ArcCatalog as ordinary tables. The table's records define which individual rasters are included in the catalog. In Geography view, the rasters included in the raster catalog are drawn in order from the first to the last record in the table. The rasters may be different in format. Any table format can be used to define a raster catalog including text files and geodatabase tables; the only requirement is that the table's columns must be NAME, XMIN, YMIN, XMAX, and YMAX. The name column contains the name or path of a raster, and the remaining columns describe its extent.

Folders can also contain TIN datasets, which contain points with x, y, and z values and a series of edges joining these points to form triangles. The resulting triangular mosaic forms a continuous faceted surface that can be used to analyze and display terrain and other types of surfaces. Either TINs or rasters can be used to represent surfaces.

For detailed information about working with rasters, please see 'Working with rasters' in the online Help system.

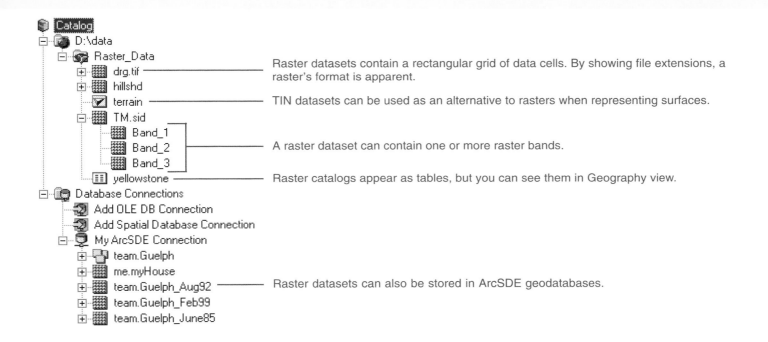

Raster datasets contain a rectangular grid of data cells. By showing file extensions, a raster's format is apparent.

TIN datasets can be used as an alternative to rasters when representing surfaces.

A raster dataset can contain one or more raster bands.

Raster catalogs appear as tables, but you can see them in Geography view.

Raster datasets can also be stored in ArcSDE geodatabases.

CAD drawings

You can access *CAD drawings* directly in ArcCatalog. CAD drawings typically have many layers, each of which represents a different type of geographic feature. For example, the drawing might contain different line layers for streets, water mains, and parcel boundaries. Other layers in the drawing may represent geographic features with points or polygons or contain annotation for labeling the features.

For each CAD drawing on disk, there is both a CAD dataset item and a CAD drawing item in the Catalog tree. The *CAD dataset* contains point, line, and polygon feature classes. The line feature class, for example, represents all line features in all layers in the drawing and their attributes. Additional feature attributes may be stored in separate tables.

CAD feature classes are drawn in Geography view using the same default symbology used by ArcCatalog when drawing shapefiles

and coverages. If you create a layer file from a CAD feature class, you can change the symbology used to draw its features, join attributes stored in separate tables to them, and select which features to display according to their attribute values. You can only analyze and edit features in ArcMap when the layer references a CAD feature class.

In the Catalog tree you'll also find a CAD drawing item, which represents all features in all layers of the drawing. In Geography view, the CAD drawing item's features appear with the symbology defined in the drawing itself. When a layer is created from the CAD drawing item, you can choose which of the drawing's layers it represents. For example, you might want to see only the streets, traffic lights, and street names on your map.

For a list of the CAD formats supported by ArcCatalog, see the online Help topic 'CAD drawings and datasets'.

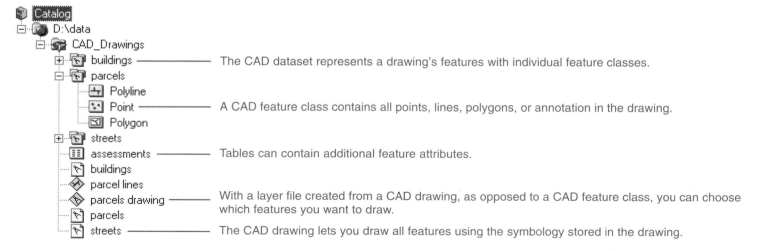

The CAD dataset represents a drawing's features with individual feature classes.

A CAD feature class contains all points, lines, polygons, or annotation in the drawing.

Tables can contain additional feature attributes.

With a layer file created from a CAD drawing, as opposed to a CAD feature class, you can choose which features you want to draw.

The CAD drawing lets you draw all features using the symbology stored in the drawing.

VPF data

The *Vector Product Format (VPF)* is a U.S. Department of Defense military standard that defines a standard format, structure, and organization for large geographic databases. VPF data is read-only in ArcCatalog. However, you can create Catalog-style metadata if you have write permission where the data is located. There are four levels of VPF data.

A VPF database is a collection of data that is managed as a unit. A VPF library, similar to a LIBRARIAN library, is a collection of coverages that fall within a defined extent and use the same coordinate system. A VPF coverage, similar to an ArcInfo coverage, may contain many feature classes. The name of a VPF coverage is the library name followed by the coverage name. For example, a coverage named "elev" in a library named "algiers" would appear in the Catalog as "algiers:elev". Coverage names are often specified in the VPF product specification.

A VPF feature class is a collection of features (primitives) that have the same attributes. Each feature class contains point (node), line (edge), polygon (face), or annotation features and has an associated feature attribute table. The feature classes within a VPF coverage represent different types of features. For example, a hydrology coverage may have feature classes representing dams, ditches, lakes, and rivers.

A coverage's features appear continuous even though they may be tiled. They must also interconnect in a manner defined by the coverage's topology. There are four levels of topology for VPF coverages (0, 1, 2, and 3). Level 0 coverages have no toplogical information. Level 3 coverages have full polygon topology.

VPF tables describe the contents of databases, libraries, coverages, and feature classes. They reside within the folder corresponding to each level of data. Tables describing the database appear below its list of coverages. Tables describing a library reside within its folder along with one folder for each coverage. In turn, a coverage's folder contains tables describing its contents and one folder for each tile, if appropriate.

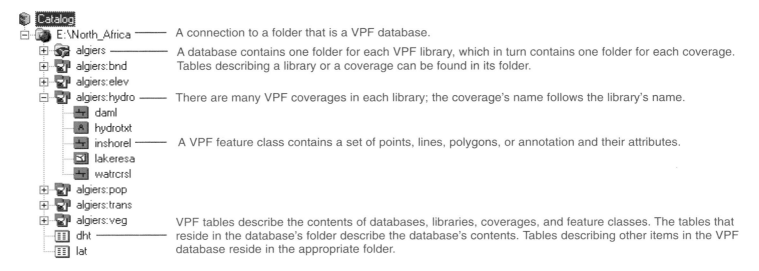

XML documents

XML is similar to HTML. HTML defines both the data and how it's presented. An HTML file contains text and tags, which tell a browser how to present the data; for example, <B>24</B> displays the text "24" in a bold font. XML, on the other hand, lets you define data using tags that add meaning. For example, <price>24</price> declares the value "24" to be a price. Other tags might declare other numbers to be totals or quantities.

You can access XML documents in ArcCatalog. You can see their contents in the Metadata tab using the XML stylesheet. *Stylesheets*, which are created using XSL, define how XML data is presented. XSL is a defined set of XML tags that can be used to query and evaluate XML data. A stylesheet retrieves values from selected elements, applies functions and formatting, and specifies how to present the resulting data. Stylesheets are similar to queries, which select and order values from tables in an RDBMS.

Because the presentation information is stored separately, you can display the same data in many different ways by changing the current stylesheet. For example, one stylesheet can present the number "24" as the formatted string "$24.00". Another stylesheet might convert the value to British pounds and present it as blue text.

ArcCatalog can create metadata for most items that appear in the Catalog tree. Metadata created by ArcCatalog is stored as XML data either in a file alongside the data source or within a geodatabase. In addition to the properties of a data source, the metadata may include documentation about the origins of the data source and how it should be used.

Metadata becomes a part of the data source itself. It is automatically moved, copied, and deleted along with the data source. Metadata XML documents won't appear in the Catalog unless they become separated from the data source.

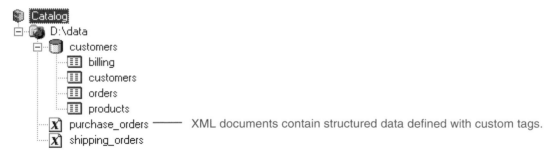

XML documents contain structured data defined with custom tags.

Internet servers

The *Internet Servers folder* lets you manage connections to ArcIMS Internet servers such as *www.geographynetwork.com*. Each *Internet server* provides access to many services. Secure services are available to users who provide the correct login information.

An *ArcIMS Image Service* is a raster representation of a complete map. After adding an image service to a map, you can turn off specific layers in the service so you only see those that are of interest to you. For example, an image service might include layers showing elevation, vegetation cover, political boundaries, mountain peaks, and cities in South America. If you were trying to determine which types of plants grow at a given elevation, you might turn off all layers except elevation and vegetation.

An *ArcIMS Feature Service* is similar to a feature dataset, which contains many feature classes. Each *ArcIMS Feature Class* represents different types of features. For example, a feature service might provide transportation data for a specific area. Its feature classes might represent streets, highways, railroads, and ferries. In ArcMap, you can analyze features in an ArcIMS feature class the same way you would feature classes in a geodatabase feature class.

Occasionally, an Internet server may be unavailable. When you lose your connection, a little red 'x' will appear on the Internet server's icon. Click the Internet server in the Catalog tree to reestablish your connection.

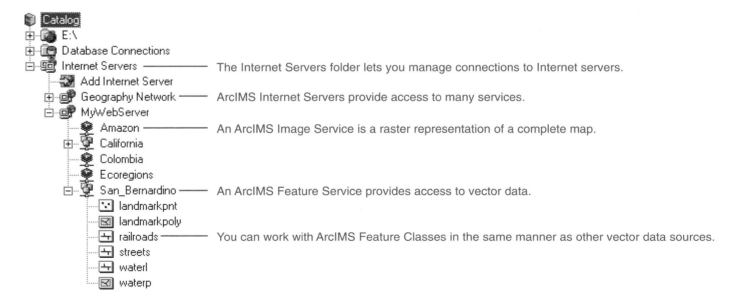

- Catalog
  - E:\
  - Database Connections
  - Internet Servers ——— The Internet Servers folder lets you manage connections to Internet servers.
    - Add Internet Server
    - Geography Network ——— ArcIMS Internet Servers provide access to many services.
    - MyWebServer
      - Amazon ——— An ArcIMS Image Service is a raster representation of a complete map.
      - California
      - Colombia
      - Ecoregions
      - San_Bernardino ——— An ArcIMS Feature Service provides access to vector data.
        - landmarkpnt
        - landmarkpoly
        - railroads ——— You can work with ArcIMS Feature Classes in the same manner as other vector data sources.
        - streets
        - waterl
        - waterp

Geocoding services

A *geocoding service* lets you convert textual descriptions of locations into geographic features. Different styles of geocoding services let you geocode different types of addresses. For example, the addresses in a table may or may not include ZIP Code information.

Geocoding services use *reference data* to find locations. A geocoding service's reference data might be a street centerlines shapefile with information about the address ranges for each section of the street. Using that service, you could take a table containing customer addresses and create a point feature class

representing the location of your customers. To geocode a table of addresses, right-click the table and click Geocode Addresses.

The *Geocoding Services folder* at the top level of the Catalog tree lets you manage existing and create new geocoding services on your computer. Geocoding services can also be stored within an ArcSDE geodatabase where they can be used by many people in your organization. You may also find ArcView GIS 3 geocoding indexes in folders on disk with the shapefile or coverage data upon which they are based; you can use these geocoding indexes as well.

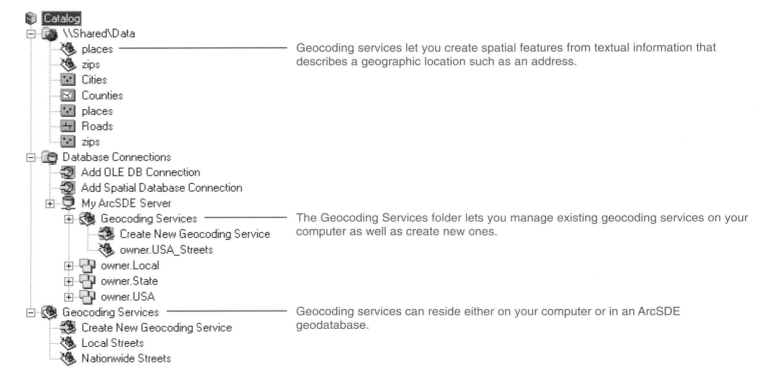

Geocoding services let you create spatial features from textual information that describes a geographic location such as an address.

The Geocoding Services folder lets you manage existing geocoding services on your computer as well as create new ones.

Geocoding services can reside either on your computer or in an ArcSDE geodatabase.

Search results

The *Search Results folder* contains your previous searches. A *search* is comprised of the name, type, location, spatial, temporal, and keyword criteria that together describe the data you want to find. For example, you can search for raster datasets for the San Francisco area that were created within the last five years by a specific agency.

When you click Find Now in the Search dialog box, your search is saved in the Search Results folder. As the Catalog finds items that satisfy your criteria, shortcuts to those items are added to your search's results list. A *shortcut* provides a link to an item that resides elsewhere.

You can modify the search's criteria, if you wish, by opening its Properties dialog box. When you have changed the desired properties and click Find Now, the search's results will be updated. To see a search's list of criteria, look at its metadata.

When you select a shortcut in the Catalog tree, you can preview the item's data and metadata. You can drag a shortcut and drop it on a map and even modify the item's properties. When you delete a shortcut you delete the shortcut itself, not the actual item. To select the item itself in the Catalog tree, right-click the shortcut and click Go To Target.

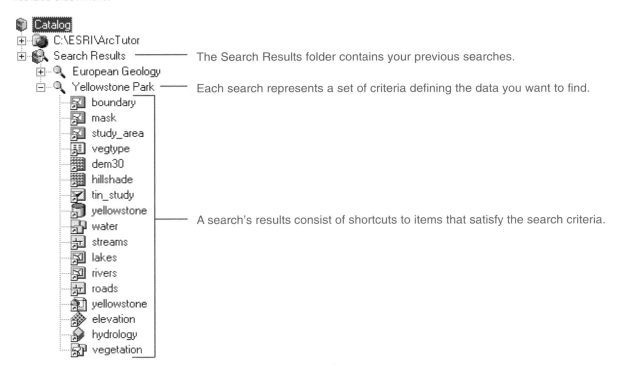

The Search Results folder contains your previous searches.

Each search represents a set of criteria defining the data you want to find.

A search's results consist of shortcuts to items that satisfy the search criteria.

Coordinate systems

There are two types of coordinate systems: geographic and projected. Geographic coordinate systems use latitude and longitude coordinates on a spherical model of the earth's surface. Projected coordinate systems use a mathematical conversion to transform latitude and longitude coordinates that fall on the earth's three-dimensional surface to a two-dimensional surface.

Each data source stores the parameters that define its coordinate system as an integral part of its data. However, those parameters can also be stored in separate files so that they can be reused when defining data for different projects.

The Coordinate Systems folder provides a location where you can organize coordinate system files. It contains many commonly used coordinate systems that are provided with ArcCatalog. You may have your own custom coordinate systems in other folders that were created with ArcInfo Workstation or with ArcCatalog. If you choose to, you can move those files into the Coordinate Systems folder.

By default, the Coordinate Systems folder is hidden in the Catalog tree. To show or hide the Coordinate System folder, check or uncheck its entry on the General tab in the Catalog's Options dialog box.

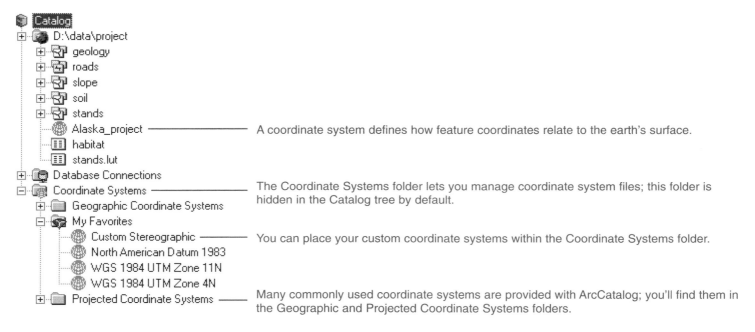

A coordinate system defines how feature coordinates relate to the earth's surface.

The Coordinate Systems folder lets you manage coordinate system files; this folder is hidden in the Catalog tree by default.

You can place your custom coordinate systems within the Coordinate Systems folder.

Many commonly used coordinate systems are provided with ArcCatalog; you'll find them in the Geographic and Projected Coordinate Systems folders.

Building the Catalog

5

When you first open ArcCatalog, you see folder connections that let you access data on your computer's local disks. Often, however, the data you use isn't stored on your computer. You can build your own catalog of geographic data by connecting to other disks or folders on the network and adding connections to databases and Internet servers. You can also add file types and hide items that you don't need for the moment.

When all your data is in one place, you gain more than just an inventory. It becomes easier to find the data. And because you can work with all types of geographic data sources the same way, regardless of their format, your data becomes easier to manage.

Adding folder connections

When you first start ArcCatalog, the Catalog tree has entries for your computer's hard disks. To access data stored on a CD, floppy disk, or another computer on the network, you must add connections to those locations. A folder connection can point to any folder to which you have access. If only one folder on your computer's C:\ drive, called "data", contains GIS data, you don't have to include the entire C:\ drive in the Catalog tree. Add a new connection that points directly to the C:\data folder, then remove the C:\ drive connection from the Catalog tree. A quick way to do this is to drag the C:\data folder from the Contents tab and drop it on the Catalog at the top of the Catalog tree. To quickly connect to folders that aren't currently available in ArcCatalog, type the folder's path into the Location text box and press Enter. If a shared folder is unavailable on the network the next time you start ArcCatalog, a little red 'x' will appear on the folder connection's icon indicating that it is unavailable. The 'x' disappears when the connection is reestablished.

Connecting to a folder

1. Click the Connect To Folder button.

2. Navigate to the folder or disk that you want to add to the Catalog.

3. Click OK.

   A new folder connection appears in the Catalog tree.

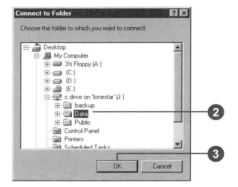

A new folder connection appears in the Catalog tree.

Disconnecting from a folder

1. Click the folder connection that you want to remove from the Catalog.

2. Click the Disconnect From Folder button.

   The folder connection is removed from the Catalog tree.

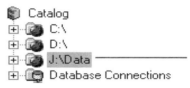

Adding spatial database connections

With ArcCatalog, you can explore and manage geographic data stored in an RDBMS through *ArcSDE*. Similarly, *SDE for Coverages* lets you access coverage, ArcInfo LIBRARIAN, and ArcStorm databases the same way you access data from an RDBMS. To access these spatial databases, you must add a connection to the Catalog.

You aren't required to type your username and password to create a connection; if you don't, you will be prompted to enter them when a connection is established. However, you can type your username and password and create a test connection if you're unsure whether the information you have is correct. If the connection test fails, contact the database administrator to ensure the database is operational. You can still add this connection to the Catalog but will be unable to retrieve data until the problem is resolved. After creating a test connection, uncheck Save name and password if you prefer not to save your login information as part of the connection. Choosing not to save login ▶

Connecting to a spatial database

1. Click the Database Connections folder in the Catalog tree.

2. Double-click Add Spatial Database Connection.

3. Type the name or IP address of the server to which you want to connect.

4. Type the name or port number of the service to which you want to connect.

5. If the data is stored in a Sybase®, SQL Server™, IBM DB2, or Informix RDBMS, type the name of the database to which you want to connect.

   If the data is stored in another RDBMS, skip this step.

6. Type your username and password for accessing the data.

7. Click Test Connection.

   If the test was successful, the button becomes unavailable. If the test fails, you won't be able to retrieve data until you've provided the correct information or the database problem has been resolved.

8. Uncheck Save Name/ Password if you don't want ▶

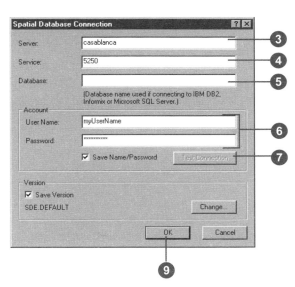

information as part of the connection can help maintain the security of the database. To make it easier to log in each time, create a connection for which you only have to enter your password each time; provide the server information and your username, then click OK.

Geodatabases managed with ArcSDE 8 can be versioned. Feature editing in ArcMap requires a versioned geodatabase. New spatial database connections will automatically access the default version. To connect to a specific version, type your server and login information and click Change. The Catalog connects to the geodatabase and lists the versions that are available. You can also choose not to save any version information as part of the connection; you might do this if you work with several different versions of the same database.

If you prefer, rather than using ArcSDE, you can connect to your spatial database using the direct connect drivers provided with ArcGIS software. To do so, you must configure your computer and provide specialized login information. For more information, see the configuration and tuning guide for your RDBMS.

this information saved with the connection.

9. Click OK.

10. Type a new name for the database connection.

11. Press Enter.

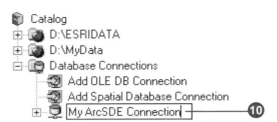

Connecting to a specific version of the database

1. Follow steps 1 through 8 for connecting to a spatial database.

   You must provide complete connection information for the database including your username and password.

2. Click Change.

3. Click the Version dropdown arrow and click the version of the database that you want to access.

4. Click OK.

5. Uncheck Save Version if you don't want to connect to the same version of the database each time.

6. Click OK.

7. Type a new name for the database connection.

8. Press Enter.

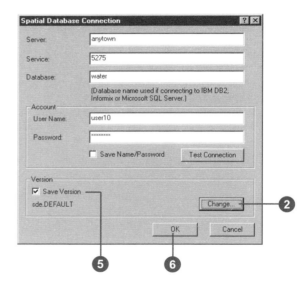

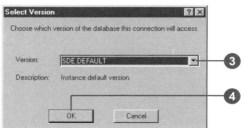

Adding OLE DB connections

You can use *Object Linking and Embedding Database (OLE DB)* providers to retrieve data from a database. The Catalog communicates with all OLE DB providers in the same way, with each provider in turn communicating with a different database. This standard lets you work with data from any database in the Catalog in the same way. Looking in an OLE DB connection, you'll see a list of tables in the database. If the provider can retrieve spatial data in *OpenGIS* format, you may also see feature classes.

Microsoft's OLE DB providers come with ArcCatalog. They let you access Jet (Microsoft Access), SQL Server™, and Oracle® databases. Another provider communicates with *open database communication (ODBC)* drivers. Additional OLE DB providers may be available from other sources. The Connection tab is different for each provider, although they all require similar information—the database to which you want to connect and your username and password. Each provider should have a Help topic with specific information on how to create a connection.

1. Double-click Database Connections in the Catalog tree.

2. Double-click Add OLE DB Connection.

3. Click the OLE DB provider you want to use for accessing data.

4. Click Next or click the Connection tab.

5. Provide the required connection information.

   The Connection tab is different for each provider; most require that you identify the database to which you want to connect and type your username and password. Click the Help button for assistance.

6. Click Test Connection.

7. Click OK if the connection test was successful.

8. Type a new name for the connection.

9. Press Enter.

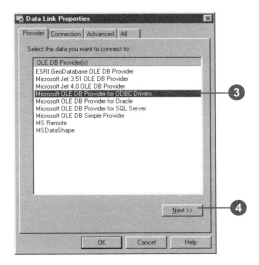

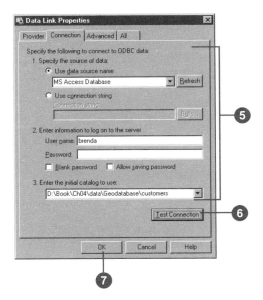

Working with database connections

Database connections are either connected or disconnected. All connections are disconnected when you start ArcCatalog. The first time a connection is selected, the Catalog attempts to connect to the database. If your login or version information isn't saved with the connection, you'll be prompted for it. When a connection is established, you can access the database's contents. Database connections remain connected until ArcCatalog is closed or you can disconnect manually. You could make a local copy of the data and disconnect before editing it onsite. The strategy you choose for connecting and disconnecting affects the number of licenses available to others who must access the database.

If the database is moved to a new computer, or the usernames and passwords for accessing its contents change, you must update your database connections, as well as the source information for layers that access data in that database, with the new connection information.

Connecting to a database

1. Click the database connection you want to use.

2. If your login information isn't saved with the database connection, a login dialog box will appear. Type the required information, then click OK.

   The red 'x' disappears from the database connection's icon.

When connected, a database connection's icon doesn't have a red 'x'.

Disconnecting from a database

1. Right-click the database connection that you want to disconnect.

2. Click Disconnect.

   A red 'x' appears on the database connection's icon.

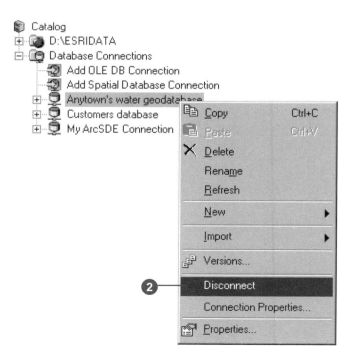

Tip

Working with Internet servers

You work with Internet servers in the same way that you work with database connections. Right-click the Internet server, then click Connect or Disconnect as appropriate. To modify your Internet server connection, open its Properties dialog box, then change your username and password for accessing secure services, for example.

Repairing a database connection

1. Right-click the database connection you want to fix and click Connection Properties.

   The appropriate dialog box appears for spatial or OLE DB database connections.

2. Change the connection properties. For example, change your username and password.

3. Click OK.

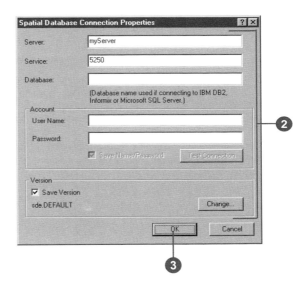

Connecting to Internet servers

To access services provided by an ArcIMS Internet server, add an Internet server connection to the Catalog. To establish the connection, provide the *Universal Resource Locator (URL)*, which uniquely identifies the server on the Internet. Your connection can either access all the services provided or you can choose the specific services in which you are interested. In addition to the services that are freely available, you may have permission to access secure services. To include these in the list of available services, you must first provide login information.

On starting ArcCatalog, all Internet servers are disconnected; a little red 'x' will appear on its icon indicating that it is unavailable. The first time an Internet server is selected, the Catalog connects to it. If your login information isn't saved with the connection, you will be prompted for it. If the connection attempt is successful, the 'x' disappears and you can access the available services.

Adding an Internet server

1. Double-click Internet Servers and double-click Add Internet Server.

2. Type the URL of the Internet server to which you want to connect. For example, type "http://www.geographynetwork.com".

3. If you don't want to connect to all services, click Just the following service(s) and click Get List. A list of all services that are freely available on the server appears below, and Get List is dimmed out. Check the services you want to use.

4. Click OK.

5. Type a new name for the Internet server.

6. Press Enter.

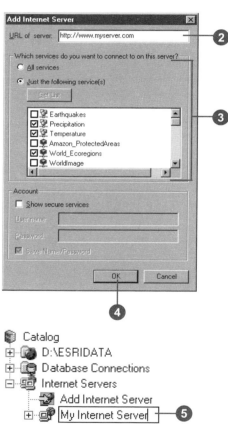

Accessing secure services

1. Double-click Internet Servers and double-click Add Internet Server.

2. Type the URL of the Internet server to which you want to connect.

3. Check Show secure services.

4. Type the username and password for accessing the secure services that are available to you.

5. If you don't want to connect to all services, click Just the following service(s) and click Get List.

   The list of services that appears below includes all free services provided by the Internet server as well as the secure services that you have permission to access.

6. Check the services you want to use.

   Uncheck the services you don't want to use.

7. Click OK.

8. Type a new name for the Internet server.

9. Press Enter.

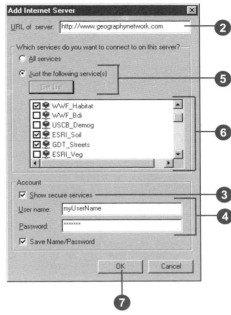

Hiding folders and items

The Catalog lets you work with data sources in many different formats and has folders that help you manage them. You can customize the Catalog to show only the folders and items that you want to work with. For example, you might only want to see the shapefiles in a folder, not the coverages and CAD drawings. When you first start ArcCatalog, the Database Connections, Internet Servers, Geocoding Services, and Search Results folders are visible. If you don't use data stored in a remote database or provided by an Internet server, you can hide those folders. If you don't use geocoding services or the Catalog's Search tool, you can hide those folders as well. Similarly, show the Coordinate Systems folder when you want to modify its contents. Check the items you want to see and uncheck those you want to hide in the General tab in the Catalog's Options dialog box.

1. Click the Tools menu and click Options.

2. Click the General tab.

3. Uncheck the items and folders you want to hide in the Catalog.

   Check the items and folders you want to show in the Catalog.

4. Click OK.

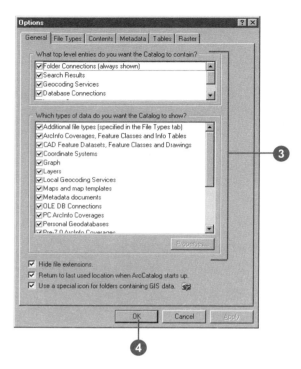

Changing an item's properties

Some items let you set properties that control how they behave in the Catalog. For example, by default the text files you see in the Catalog are files stored on disk that have the file extensions .txt, .csv, .tab, or .asc. These file extensions are commonly used to denote files that contain American Standard Code for Information Interchange (ASCII) data. You can change the properties of text files and control which ones will appear in ArcCatalog. If files with the .csv extension are the only ones on your computer that describe geographic features, you might want only those files to appear in the Catalog as text files.

1. Click the Tools menu and click Options.

2. Click the General tab.

3. Click the item whose properties you want to change.

4. Click Properties.

5. Modify the appropriate properties for this item.

6. Click OK.

7. Click OK.

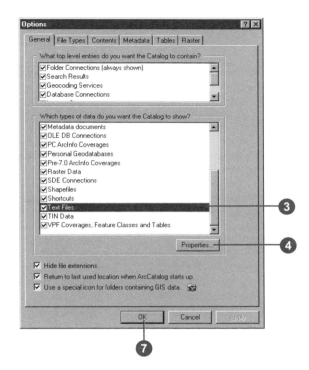

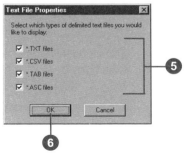

Choosing which raster formats appear in the Catalog

The raster formats that can be directly accessed by ArcCatalog are listed in the Raster tab of the Options dialog box. At times, you may not want to see all types of rasters in the Catalog. For example, the same data may be stored in different formats and you want to work with only one of those formats at a time. Raster files often have a well-known extension describing the format of the data they contain—for example, .jpg. If a raster's data is in a format supported by the Catalog but its file extension doesn't appear in the Raster tab's list, it won't appear in the Catalog; for example, only files with the extension .tif are recognized as TIFF images by default and not .tiff files. You can customize the list of file extensions that are associated with a raster format.

Not all rasters have file extensions, and some are stored as folders. Because identifying such rasters takes more time, the Catalog skips those without extensions by default when searching through a folder's ▶

Displaying specific raster formats

1. Click the Tools menu and click Options.

2. Click the Raster tab.

3. Click the option to Search only files that match the following file extensions to find valid raster formats.

4. Uncheck the formats you don't want to see as rasters in the Catalog.

   Check the formats you want to see in the Catalog.

5. Click OK.

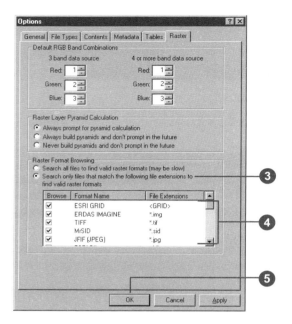

Displaying all raster formats

1. Click the Tools menu and click Options.

2. Click the Raster tab.

3. Click Search all files to find valid raster formats.

4. Click OK.

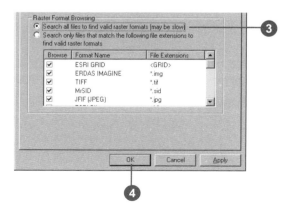

contents. Therefore, they won't appear in the Catalog. The exception is the ESRI grid format, which is quick to search for even without a filename extension. By default, the Catalog will always search for grids. If you choose to display all rasters, the contents list will be slower to appear, but you will see an accurate listing of all rasters in all folders.

Changing a raster format's file extensions

1. Click the Tools menu and click Options.

2. Click the Raster tab.

3. Scroll down until you see the appropriate raster format in the list.

4. Click in the File Extensions column to the right of the raster format's name.

5. Move the cursor to the end of the file extensions list.

6. Type a semicolon (;) and then type the letters that comprise the file extension.

   For example, if you add the file extension "jpeg" into the File Extensions column to the right of JFIF (JPEG), both .jpg and .jpeg files will be recognized as JFIF images in ArcCatalog.

7. Press Return.

8. Click OK.

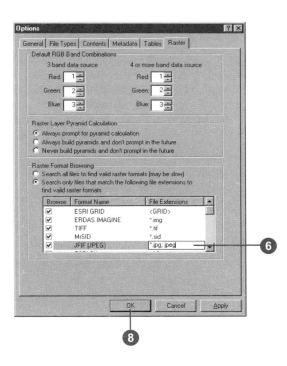

Adding file types

Many files that you would see in a folder using Windows Explorer aren't initially visible in the Catalog. Some of these files may contain information that you need when working with geographic data. To see them, you must add their types to the Catalog's file types list.

Create your own file type by defining the file extension, description, and the icon you want to use to represent those files. Files with that extension will appear in the Catalog with that icon, and the description will appear in the Type column in Details view. For example, to see ArcView GIS 3 project files in the Catalog, add the file extension "apr" to the file types list.

Some of the files you want to see in the Catalog may already be registered with the operating system such as Microsoft Word documents. You can add file type information for these files to the Catalog using the information in the registry. When you double-click a file whose type is registered with the operating system, the Catalog will open it in the appropriate application.

Defining a file type

1. Click the Tools menu and click Options.

2. Click the File Types tab.

3. Click New Type.

4. Type the file extension.

5. Type a description of the new file type.

6. Click Change Icon.

7. Click an icon.

   Or click Browse, then navigate to and click the application whose icon you want to use for the file type, and click Open.

8. Click OK in the Change Icon dialog box.

9. Click OK.

   The new file type is added to the list.

10. Click OK.

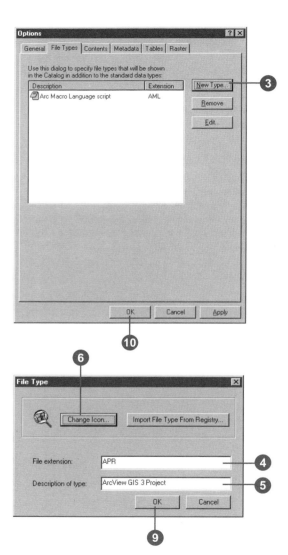

Editing a file type

Edit the properties of a file type to change the description that appears in Details view or to change the icon used to represent it in the Catalog. Select the file type in the File Types tab of the Options dialog box and then click Edit. Modify the appropriate properties, click OK, then click OK in the Options dialog box.

Importing a file type

1. Click the Tools menu and click Options.

2. Click the File Types tab.

3. Click New Type.

4. Click Import File Type From Registry.

5. Scroll down until you see the file type you want to use and click it in the list.

6. Click OK.

   The file type's properties appear in the File Type dialog box.

7. Click OK.

   The file type is added to the list.

8. Click OK.

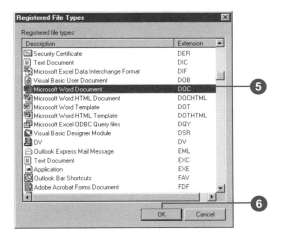

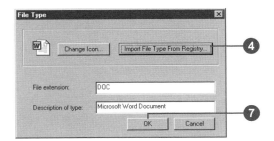

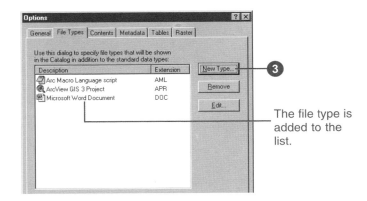

The file type is added to the list.

Managing the Catalog's contents 6

With the Catalog tree and the Contents tab you get a bird's-eye view of how your data is organized. The Catalog has many built-in functions that make managing your data easy. Copy, move, rename, and delete items in folders and databases using standard Windows shortcuts. Several wizards and tools are also available for converting your data from one format to another.

With ArcCatalog, you can quickly learn a great deal more about a data source than just its name. The Contents tab provides different ways of listing the contents of items in the Catalog. You can see thumbnail snapshots illustrating the contents of maps and data sources. Or customize the columns that appear in Details view so you can compare the properties and metadata of several data sources at a glance.

Whether you use it to browse for data to add to maps or manage your organization's ArcSDE geodatabase, ArcCatalog is bound to become a permanent feature on your desktop!

Viewing an item's contents

In the Contents tab, items that contain subsets of items—such as folders, databases, coverages, and feature datasets—appear at the top of the Contents list and are grouped by type. This behavior is similar to Windows Explorer, which also places folders at the top of its contents list. Individual items such as shapefiles, maps, and tables are listed below in one group. Even though raster datasets may contain several raster bands, they are not grouped separately at the top of the list because they can be either single-band or multiband rasters.

Listing an item's contents

When you select items such as folders or databases in the Catalog tree, the Contents tab lists the items they contain such as maps or tables. You can display the Contents list in several ways. To change its appearance, use the buttons on the Standard toolbar.

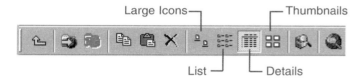

Large Icons view

In Large Icons view, each item in the Contents list is represented by a large icon. The icons are ordered from left to right in rows increasing from top to bottom.

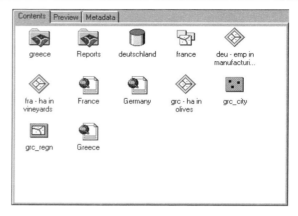

List view

In List view, each item in the Contents list is represented by a small icon. The icons are ordered from top to bottom in columns increasing from left to right. ▶

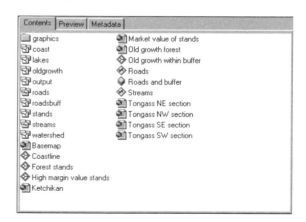

Details view

In Details view, each item in the Contents list has a small icon. The items are arranged in a long list. Properties for each item appear in columns. By default, there are only columns describing the item's Name and Type, but you can add columns representing other properties and information stored in the metadata. If a property is inappropriate, the item's value for that property is blank. For example, the file size of a dBASE table is well-known, but this property doesn't apply to tables stored in a database.

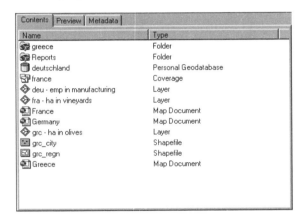

Thumbnails view

Thumbnails view displays a snapshot illustrating the contents of each item in the Contents list. A thumbnail might provide an overview of all features in a coverage or a detailed look at the features symbolized by a layer. If an item can't or doesn't have a thumbnail, its icon is displayed in a gray box.

For all items except maps, thumbnails are created and updated manually in Geography view. A map's thumbnail is created automatically when it is saved.

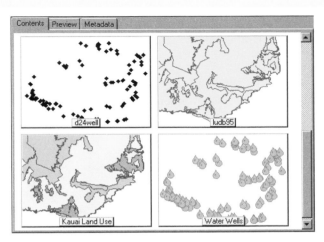

Viewing the contents of individual items

Items such as maps, shapefiles, and tables don't contain other items. When you select individual items, the Contents tab lists the properties and metadata that you would see in Details view as well as the item's thumbnail.

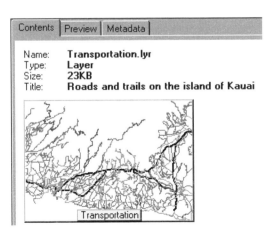

Working in Details view

Sometimes you need information about an item before you can decide whether or not to use it. Details view can show you a few properties and metadata elements for the items in the list so you can quickly see some of the differences between them. To better see the values for a property, resize a column. To compare the items in the list, sort them by a property's values.

Changing a column's width

1. Click the item whose contents you want to list in the Catalog tree.

2. Click the Contents tab.

3. Click the Details button on the Standard toolbar.

4. Position the mouse over the edge of the column you want to resize.

   The pointer's icon changes.

5. Drag the column's edge to the desired width and drop it.

   The column is resized.

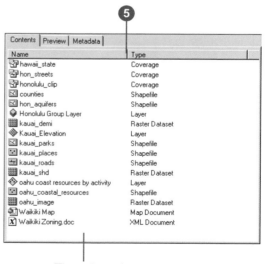

The column has been resized.

Sorting the Contents list by a property

1. Click the item whose contents you want to list in the Catalog tree.

2. Click the Contents tab.

3. Click the Details button on the Standard toolbar.

4. Click the heading of the column whose values will be used to sort the list.

   The first time you click the column heading, its values are sorted in ascending order.

5. Click the column heading again to sort the list in descending order.

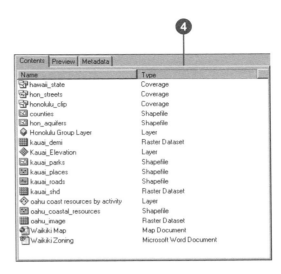

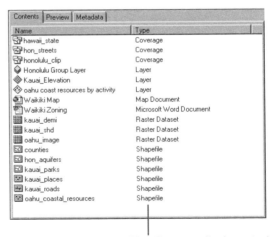

The Contents list is sorted by the property's values.

Changing the columns you see

When you start using ArcCatalog, you'll see two columns in Details view that show the Name and Type of each item. You can add columns that show additional properties or information stored in the item's metadata. Check the columns you want to see in the Contents tab in the Options dialog box. When an individual item is selected in the Catalog tree, the values that would appear for that item in all columns in Details view are listed in the Contents tab. A column may not apply to all items. For example, the Size and Modified properties in the standard columns list apply to maps and layers, but no values will appear in these columns for data stored in a database. Values won't appear in metadata columns if metadata hasn't been created for an item or if the element it corresponds to is empty.

You can add your own property and metadata columns to the lists in the Options dialog box. To do so, you must type the exact name of the property or metadata element into the Property text box in the Add Standard Column or Add ►

Showing and hiding standard and metadata columns

1. Click the Tools menu and click Options.

2. Click the Contents tab.

3. Check the properties and metadata elements you want to show in Details view.

   Uncheck the properties and metadata elements you want to hide in Details view.

4. Click OK.

   All the checked properties and metadata elements appear in Details view and Individual Item view.

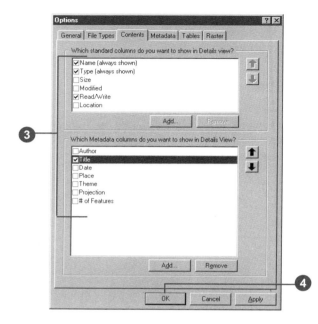

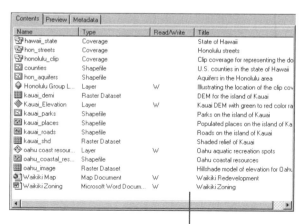

The checked properties and metadata elements appear in Details view and Individual Item view.

Metadata Column dialog box. Into the Caption text box, type the name of the column as it should appear in Details view.

The standard columns list includes all properties defined in ArcGIS software that can appear in Details view. However, if you use custom GxObjects, additional properties may be available to you. To add them to the list, you must provide the exact name of the property as it is defined in the GxObject itself; for example, the property associated with the Size column is named "ESRI_GxObject_FileSize".

You might add additional elements to the metadata columns list to compare the percent cloud cover for several remotely sensed images, for example. To add a metadata column such as cloud cover, you must provide the element's "path". An element's path describes how to navigate the hierarchy of the metadata XML document from its root element to find the element you want to display. For example, in the FGDC's *Content Standard for Digital Geospatial Metadata (CSDGM)*, the Percentage Cloud Cover element is contained in the Data Quality section of the XML document. The XML element corresponding to the metadata element is ▶

Adding columns to the list

1. Click the Tools menu and click Options.

2. Click the Contents tab.

3. Under the list of standard or metadata columns, click Add.

4. In the Caption text box, type the column name as it should appear in Details view. In the Property text box, type the name of the property or the metadata element's path.

   Or click the Caption or Property dropdown arrow and click the property or metadata element that you want to add to the list.

5. If you wish, type an appropriate default width for this column in pixels.

6. Click OK.

   The new column is added to the bottom of the appropriate list.

7. Click OK.

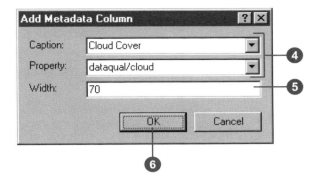

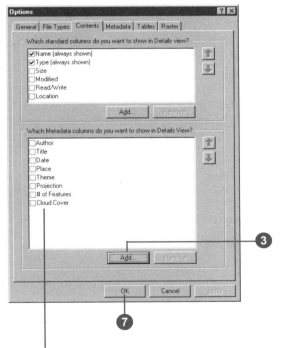

The new column is added to the appropriate list.

its short name as defined in the CSDGM—"cloud" and "dataqual", respectively. The dataqual element is contained in the root element of the XML document, "metadata". To add this Cloud Cover element as a column in Details view, provide the path "dataqual/cloud" in the Property text box; the root element isn't included in the path. If you're not familiar with the CSDGM or the *ESRI Profile on FGDC Metadata,* you can determine a metadata element's path by looking at metadata with the XML stylesheet.

All columns that are listed in the Options dialog box by default appear in the Caption and Property lists in the Add Standard Column and Add Metadata Column dialog boxes. If you delete a default column you can add it again later by choosing its caption or name from either the Caption or Property dropdown lists.

Tip

Changing the order of columns

Because you can't move columns in Details view, you must control the order in which they appear in the Options dialog box. Columns will appear in order from the first standard column to the last metadata column in the list.

Rearranging columns

1. Click the Tools menu and click Options.

2. Click the Contents tab.

3. Click the property or metadata column whose position you want to change in the list.

4. Click the Up and Down arrows to change the column's position.

5. Click OK.

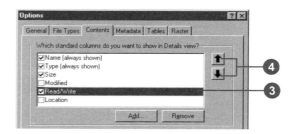

Removing columns

1. Click the Tools menu and click Options.

2. Click the Contents tab.

3. Click the column that you want to remove from the list.

4. Click Remove.

   The column no longer appears in the list.

5. Click OK.

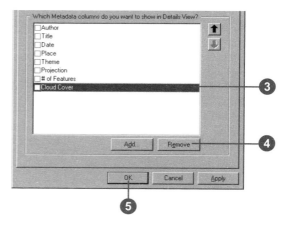

Exploring an item's properties

While very useful, Details view can only provide you with a limited amount of information. To learn more about an item, open its Properties dialog box. If you have write permission for the data source, you may be able to change some of its properties. For example, when looking at a raster's properties, you can find out if statistics have been calculated or if its coordinate system has been defined. If you have write permission, you can update these properties.

1. Right-click the item whose properties you want to see.

2. Click Properties.

3. Examine the properties using the tabs and buttons in the Properties dialog box.

4. Click Cancel to dismiss the Properties dialog box.

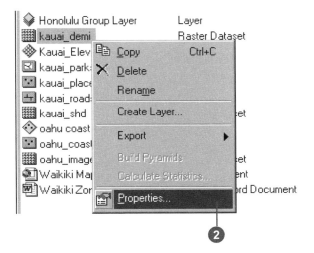

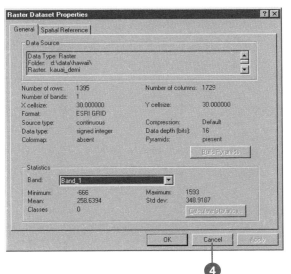

Organizing your data

With ArcCatalog, managing maps and geographic data is as easy as managing files with Windows Explorer. Use the standard Windows shortcuts and drag and drop techniques to copy, move, rename, and delete items in the Catalog. These features make it easy to organize not only data sources on disk (such as coverages and shapefiles), but data stored in databases as well.

Keep in mind that when you delete a database connection, you are deleting the connection itself, not the database or its contents. However, when you delete a personal geodatabase, you are deleting the Access database file and all the data it contains.

You can copy or move items such as database connections and coordinate systems to and from their folders in the Catalog tree and other folders on disk. For example, you can place ArcSDE geodatabase connections and custom coordinate systems in a shared folder on the network so everyone in your organization can access them.

Creating a new folder

1. Click the folder connection or folder in which you want to create a new folder.

2. Click the File menu.

3. Point to New and click Folder.

4. Type a new name for the folder.

5. Press Enter.

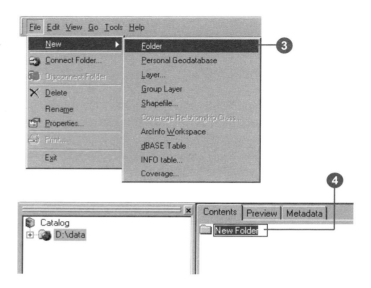

Creating a new personal geodatabase

1. Click the folder connection or folder in which you want to create a new personal geodatabase.

2. Click the File menu.

3. Point to New and click Personal Geodatabase.

4. Type a new name for the personal geodatabase.

5. Press Enter.

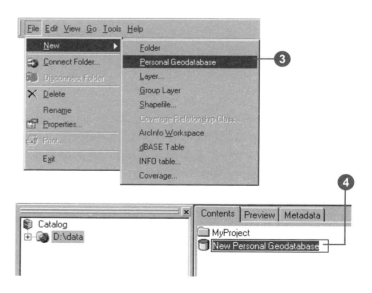

With ArcCatalog, you can rename an entire coverage and also rename region and route feature classes within a coverage.

Copying an item

1. Click the item you want to copy.

2. Click the Copy button.

3. Click the folder or geodatabase to which you want to copy the item.

4. Click the Paste button.

Renaming an item

1. Click the item you want to rename.

2. Click File and click Rename.

3. Type the new name.

4. Press Enter.

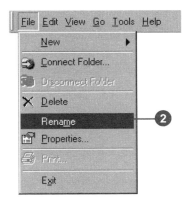

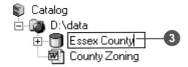

Deleting an item

1. Click the item you want to delete.

2. Click the Delete button.

Converting data to a different format

Converting data from one format to another is a common task, usually done at the beginning of a project. You might receive data in Interchange (e00) format and have to import that data to a coverage. Or you might export data from a geodatabase to a shapefile before sending it to someone else.

ArcCatalog makes it easy to change a data source's format. Right-click the data source whose format you want to change and point to Export. A list of the data converters that are appropriate for the selected data source will appear. Similarly, when a geodatabase is selected, you can import data from several different formats.

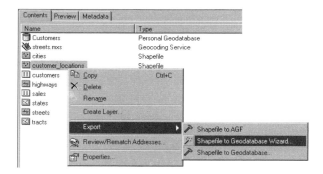

The data converters are wizards and tools that are provided with ArcToolbox. Conversion parameters will be set to suggested values based on the data source's format and the type of data it contains. However, you can change those parameters and export the data to a different coordinate system. You can also choose which attributes are exported and what their new names will be.

You can convert many items at once. For example, if you want to load several tables into a geodatabase, select the folder in which they reside in the Catalog tree. While holding down the Ctrl key, click the appropriate tables in the Contents list. Click the File menu, point to Export, then click Table to Geodatabase. The Table

To Geodatabase tool appears with a batch grid at the bottom. There will be one row in the grid for each table.

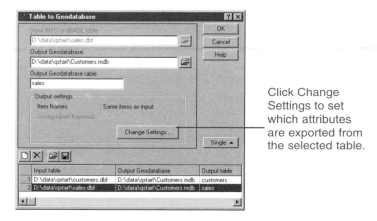

Click Change Settings to set which attributes are exported from the selected table.

You can add rows to or delete rows from the batch grid using the appropriate buttons on the tool. You can set conversion parameters for each table individually or set parameters for many tables at once. Select the appropriate rows in the batch grid before setting parameters for those tables.

Click the Help button for detailed instructions about how to use the data converters. You can also find more information about working with wizards and tools in the ArcToolbox section in the online Help system; in particular, you might find the task 'Batch processing' useful.

More data conversion tools are available in ArcToolbox than those that appear in the Catalog. In ArcToolbox, you'll also find data management and analysis tools. All ArcToolbox tools are available within the Catalog as commands that can be dropped onto a table or menu. For details, see Chapter 15, 'Customizing ArcCatalog'.

Exploring an item's geography

In ArcCatalog, you can preview an item's geographic data without first having to create a map. Who hasn't been told about a map they can use, then promptly forgotten its name even though they remember what it contains? Now you can look at a map in the Catalog to make sure it's the one you want before opening it in ArcMap.

Have you ever been almost certain that two coverages with different names actually contain the same data? You can zoom in and pan around to get a good look at their features to decide which coverage has the most recent data. ArcCatalog's Geography view lets you quickly decide which maps and data sources you want to use.

Previewing an item's geographic data

The Preview tab lets you see a selected item's geographic data. For items that contain both geographic data and tabular attributes, you can toggle between the Geography and Table views using the dropdown list at the bottom of the Preview tab.

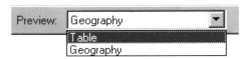

Geography view draws each feature or annotation in a vector dataset, each cell in a raster dataset, and each triangle in a TIN dataset. When you are previewing data in Geography view, the Geography toolbar is active. You can explore the selected item's geographic data using the buttons on the Geography toolbar.

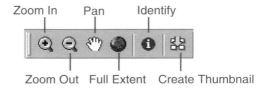

Thumbnails illustrate the contents of items containing geographic data. Once created, they can help you find the data you need quickly and easily. For all items except maps, thumbnails are created manually by clicking the Create Thumbnail button on the Geography toolbar. The thumbnail, which records exactly what you see in Geography view, is stored within the item's metadata.

How the Catalog draws geographic data

When ArcCatalog draws vector data, features are drawn using default symbology defined by the Catalog: polygons are yellow, lines are blue, points are black diamonds, and annotation is gray text. Thumbnails for vector data sources have the same appear-

ance because there is no symbology stored within the data source itself. There are two exceptions to this rule: CAD drawings and geodatabase feature classes that have subtypes.

A CAD drawing has two representations in the Catalog tree: a CAD dataset item and a CAD drawing item. The CAD dataset has point, line, and polygon feature classes. With a line feature class, you'll see all line features in the drawing. As with shapefiles, CAD feature classes are drawn with the Catalog's default symbology.

If you select the CAD drawing in the Catalog tree, each layer in the drawing is drawn with the symbology defined in the CAD drawing. Lot, building, tree, and road lines all have different symbology if they are in different layers in the drawing. You'll also see point, polygon, and annotation features. ▶

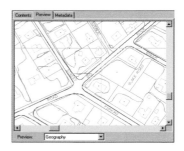

Feature classes in a geodatabase can have subtypes, which represent different categories of features. For example, lots might be residential, commercial, or agricultural. When previewing a feature class with subtypes in Geography view, each feature is symbolized according to its subtype.

A raster dataset's appearance in Geography view depends on how many raster bands it has. If there is only one raster band, the cell values are symbolized using a grayscale color ramp unless the raster dataset has a colormap. Colormaps define the color that will be used to represent a specific value.

Raster datasets with more than one raster band are displayed by creating a composite image. In the Options tab, you can specify which raster bands supply the red, green, and blue display values. For example, if a remotely sensed image had seven bands, you might want to see bands five, four, and three.

Raster catalogs appear in the Catalog tree as tables. However, ArcCatalog will draw raster catalogs in Geography view. The table defines which rasters are included in the raster catalog and the order in which they are drawn.

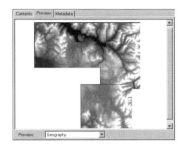

TIN datasets are drawn using the Catalog's default symbology. A triangle's color reflects its elevation. Any mass points and break lines that were used to create the TIN will be drawn as part of the TIN dataset.

Layers include a shortcut to data that's stored elsewhere and information about how to symbolize and label the data. For example, the size of a city point might reflect its population, and the symbol used to represent a city might reflect whether or not it is a capital city. When you preview a layer in the Catalog, you see it exactly as it will appear in a map. ▶

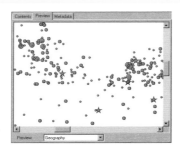

Layers may also include instructions for joining attributes stored in another table to the data, a choice of which attributes should be available in the map, and even alternate names for those attributes that are more descriptive.

Group layers combine data from many data sources; when added to a map, there is one entry in the table of contents for the group. For example, a group layer named transportation might combine highway, road, and trail coverages. Group layers can combine data sources that have different formats such as TIN datasets and shapefiles.

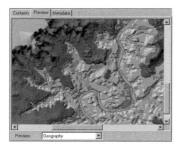

In ArcMap, you can create graphs that describe the relationships between the attributes of different features. Once created, graphs can be saved to disk. This lets you add a graph with the same format and colors to several maps. For example, the same graph

format used in different maps might compare how prevalent a group of industries are in different regions of the country. Before adding a graph to a map, you can look at its contents in Geography view.

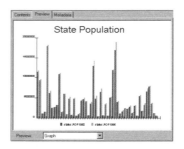

Not only can you preview individual data sources, layers, and graphs, but ArcCatalog can also draw map documents and templates in Geography view. This lets you decide whether or not you've found the correct map before opening it.

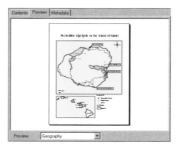

Overall, Geography view is very useful. The remaining tasks in this chapter show how to use the tools in Geography view to explore your data.

Exploring geographic data

When the selected item in the Catalog tree contains geographic data, you can preview that data without having to create a map—simply choose Geography from the Preview dropdown list on the Preview tab. In Geography view, each feature in a vector dataset, each cell in a raster dataset, or each triangle in a TIN dataset is displayed. Explore the data using the buttons on the Geography toolbar. You can zoom in, zoom out, and pan around to see different areas or reset the display to draw the entire dataset. With the Identify tool, you can click a feature, raster cell, or TIN triangle and see a list of its attributes.

Tip

Stop drawing

You can press the Esc key at any time while you are previewing geographic data to stop ArcCatalog from drawing the data.

Tip

Zooming out

Zooming out on a dataset is the exact opposite of zooming in. After clicking the Zoom Out button, click or drag a box over the dataset.

Zooming in on a data source

1. Click the Zoom In button on the Geography toolbar.

2. Drag a box over an area to see it in detail or click an area to center it in the display; it will zoom in by 10 percent.

   After you zoom in, the scale of the displayed features is larger.

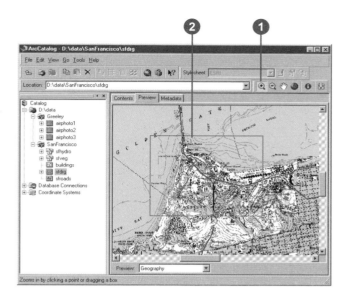

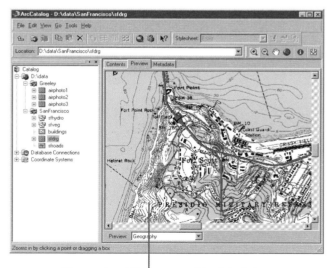

The scale of the geographic data is larger.

Panning around a data source

1. Click the Pan button on the Geography toolbar.

2. Drag the geographic data in the display and drop it in a new location.

   After panning, you'll see the features in a new geographic area.

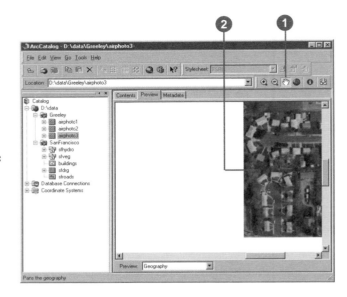

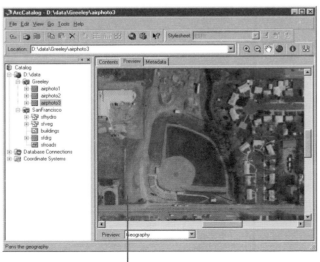

The display now shows data in a new geographic area.

Identifying geographic data

1. Click the Identify button on the Geography toolbar.

2. Click the feature, raster cell, or TIN triangle whose attributes you want to see.

   The feature flashes or the cell or triangle is singled out, and its attributes appear in the Identify Results window.

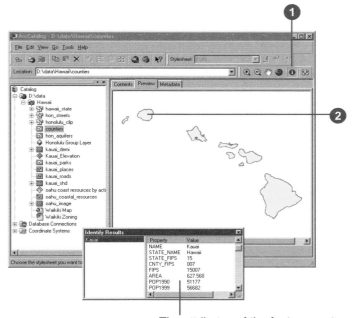

The attributes of the feature, raster cell, or triangle appear in the Identify Results window.

Creating thumbnails

Maps, layers, and data sources can have a thumbnail that appears in the Contents tab and in the item's metadata when you use the ESRI stylesheet. A thumbnail illustrates the geographic data in a data source, the features a layer represents, or a map's layout. It might show an overview of all features or a close-up illustrating a layer's symbology. Thumbnails draw quickly because they are snapshots; the Catalog doesn't draw the data itself when displaying a thumbnail.

While a map's thumbnail is generated automatically when the document is saved, you must create thumbnails for layers and data sources manually. Update a layer's thumbnail when the data or the symbology changes. A map's thumbnail is stored as part of the map document or template. Thumbnails for layers and data sources are stored in their metadata. If metadata doesn't exist before creating a thumbnail, metadata will be created automatically, but it will only contain the thumbnail. Adding properties and documentation to the metadata is a separate process.

1. In the Catalog tree, click the layer for which you want to create a thumbnail.

2. Click the Preview tab.

3. Click the Preview dropdown arrow on the Preview tab and click Geography.

4. Click the Zoom In button on the Geography toolbar and zoom to the area that best represents the layer's contents.

5. Click the Create Thumbnail button.

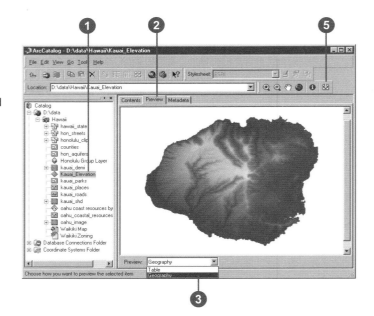

Changing raster previewing properties

When you preview a single-band raster dataset, the value of each cell is drawn as a color or a shade of gray depending on the data. When you preview a multiband raster dataset, three of its bands are combined to form a composite image, in which each band supplies either the red, green, or blue display value. From the Raster tab in the Options dialog box, you can choose which band will provide which value. You can specify a different set of defaults, mapping bands to red, green, and blue (RGB) display values for datasets with three or more bands.

1. Click the Tools menu and click Options.

2. Click the Raster tab.

3. Under 3 band data source, type the number of the band that will provide the red display values.

   Type the number of the band that will provide the green values.

   Type the number of the band that will provide the blue values.

4. Repeat step 3 for raster datasets with four or more bands.

5. Click OK.

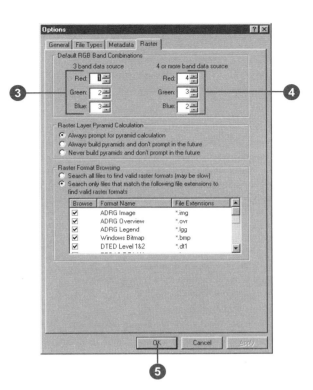

Creating raster pyramids

The amount of detail you see when drawing a raster depends on its cell size. If each cell covers a very small area so that details are maintained, then more cells are required to cover the same region: more detailed rasters will occupy more space on disk and take longer to draw. To avoid this problem, you can create pyramids, where the original data has several levels of resolution. With pyramids, the coarsest level of resolution displays quickly when drawing the entire dataset. As you zoom in, levels with finer resolutions are drawn; performance is maintained because you're drawing successively smaller regions.

By default, when you select a raster in ArcCatalog that is larger than 1,024 by 1,024 cells and doesn't have pyramids, you will be asked whether or not you want to create them. If you dislike being prompted each time, you can choose to always create pyramids or never create pyramids. On building pyramids, a reduced resolution dataset (.rrd) file is created. For an uncompressed raster, this file is approximately 8 percent of the original raster file size.

1. Click the Tools menu and click Options.

2. Click the Raster tab.

3. Click the appropriate choice describing when pyramids should be created.

4. Click OK.

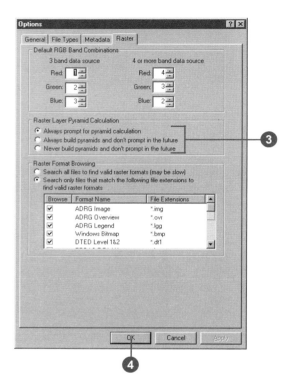

Exploring the values in a table 8

Whether you want to know if a data source has a specific attribute or if its values are correct, you can find the answer in ArcCatalog. You can sort a table's records by the values in one or more columns, get statistics describing a column's values, or locate a specific value in a table. When exploring values, you can easily add and delete attributes without having to open the data source's Properties dialog box. Table view lets you learn a lot about the contents of a table without having to create a map.

Previewing the values in a table

The Preview tab lets you explore the selected item's data in either Geography or Table view. For items that contain both geographic data and tabular attributes, you can toggle between Geography and Table view using the dropdown list at the bottom of the Preview tab. This chapter focuses on Table view.

In Table view, the table's columns and rows and the value for each cell are displayed. Explore the table's contents using the scroll bars and the buttons at the bottom of the table. Once you click inside the table you can also use the arrow keys on your keyboard to explore its contents.

Unlike ArcMap, you can't select records while exploring a table's values in ArcCatalog; you can only view all the records in a table.

When working with tables stored in databases, many people may view and edit a table's contents at the same time. If the values in the table you're previewing are changing, you may want to reload the table's contents periodically to ensure you're working with the most current values. Click Options at the bottom of the table and click Reload Cache to refresh the table's values.

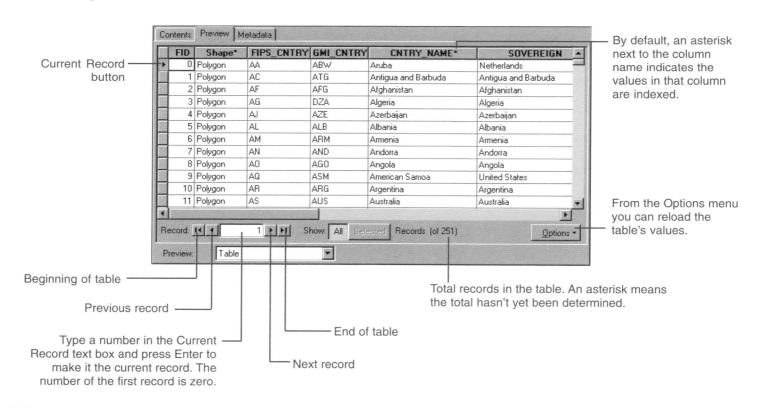

Current Record button

By default, an asterisk next to the column name indicates the values in that column are indexed.

From the Options menu you can reload the table's values.

Beginning of table

Previous record

Type a number in the Current Record text box and press Enter to make it the current record. The number of the first record is zero.

End of table

Next record

Total records in the table. An asterisk means the total hasn't yet been determined.

Changing how tables look

You can change the appearance of tables in ArcCatalog to make it easier to read their values. For example, you may want to change the size, color, and type of font used. To quickly pick out which columns are selected— for example, the columns used to sort records in a table—the background color of selected columns is set to the selection color. When you see an asterisk (*) next to a column's name, the values in that column are indexed; indexes can improve the performance of queries. Use the Options dialog box to change the font, selection color, and index character used in Table view to suit your preferences. These changes affect the way all tables appear in the Catalog.

When examining an individual table's contents, you can resize columns to better see their values. You can also reposition them to compare the values in one column to the values in another. Frozen columns are locked in position at the left of the table and are separated from the other columns by a heavy black line. When scrolling horizontally, all other columns move normally, but the frozen ▶

Setting the font, highlight color, and index character

1. Click the Tools menu and click Options.

2. Click the Tables tab.

3. Click the Selection color dropdown arrow. In the color palette, click the color you prefer to use for selected columns.

4. Click the Table Font dropdown arrow and click the font you prefer. Type the font size you want to use or pick a size from the dropdown list. If you want to set the font's color, click the font color dropdown arrow and click the text color you prefer.

5. Type the character to use when indicating whether or not a column has an index associated with it. Uncheck Show index fields with if you prefer not to see which columns are indexed.

6. Click OK.

   The appearance of all tables in the Catalog will change according to the settings in the Options dialog box.

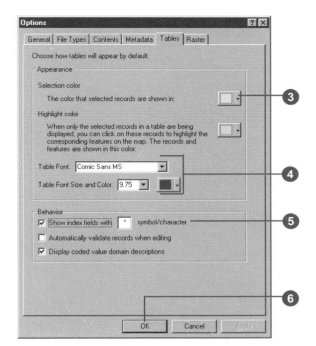

The appearance of all tables changes according to the settings in the Options dialog box.

columns remain fixed. Freezing a table's columns can be helpful when exploring its contents. Suppose a shapefile represents all counties in a state, and its attributes describe the demographics of each county. By freezing the county name column, you can easily place other attribute columns next to it and learn which county has how many households or farms. After unfreezing a column, it reappears in the table to the left of the other scrolling columns. The changes you make when resizing, repositioning, and freezing a table's columns are lost when you select a different data source in the Catalog tree.

Changing a column's width

1. Position the mouse over the right edge of the column you want to resize.

   The pointer's icon changes.

2. Drag the column's edge to the desired width.

   A black line indicates where the right edge of the column will be located.

3. Drop the edge of the column.

   The column is resized.

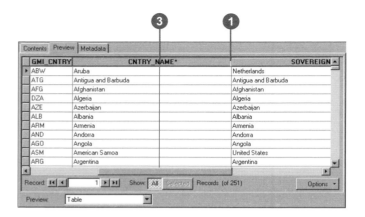

The column has been resized.

Repositioning a table's columns

1. Click the heading of the column you want to move.

2. Click the column's heading again, but hold down the mouse button.

   The pointer's icon changes.

3. Drag the column heading to where you want the column to appear.

   A red line indicates where the column will be located.

4. Drop the column.

   The column appears in the new position.

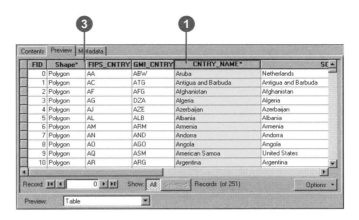

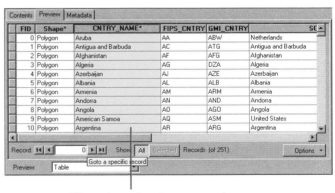

The column has been moved to a new position.

Tip

Rearranging frozen columns

After freezing more than one column, you may want to arrange them in a logical order. Not only can you rearrange columns in the main body of the table, but you can also rearrange columns within the frozen section of the table.

Tip

Unfreezing columns

It is also possible to unfreeze columns. Simply right-click the heading of the frozen column you wish to unfreeze, then click Freeze/Unfreeze Column.

Freezing a column

1. Press the Ctrl key on the keyboard and click the headings of the columns that you want to freeze.

2. Right-click the heading of one of the selected columns and click Freeze/Unfreeze Column.

   The columns are frozen.

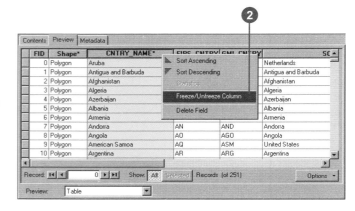

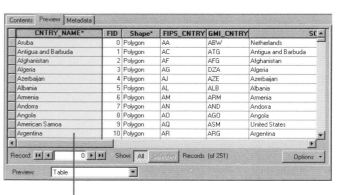

The column has been frozen.

Calculating statistics

When exploring a data source's contents, you can get statistics describing the values in numeric columns. You'll see how many values the column has as well as the sum, minimum, mean, maximum, and standard deviation of those values. A histogram is also provided showing how the column's values are distributed. Statistics are calculated for all numeric columns in the table. To see a description of another column's values, click its name in the Field dropdown list.

1. Right-click the heading of a column that contains numeric data.

2. Click Statistics.

   In the Statistics dialog box, you'll see information about the values in the column whose heading you clicked.

3. If you want to see statistics for another numeric column, click the Field dropdown arrow and click the column's name.

4. Click the Close button when you are finished exploring the statistics.

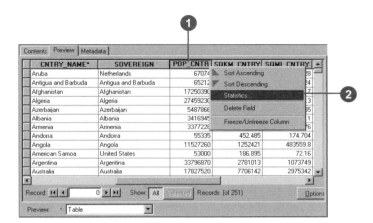

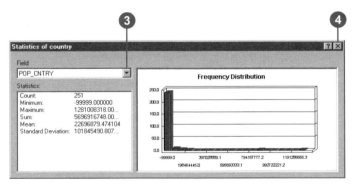

Sorting records in a table

Sorting the rows in a table lets you more easily derive information about its contents such as which county had the highest rent last year. After sorting a column's values in ascending order, the values are ordered from A to Z or from 1 to 10. With descending order, a column's values are arranged from Z to A or from 10 to 1.

Sometimes it's helpful to sort a table's rows by more than one column. For example, it might be more helpful to sort the counties first by state and then by rent—the effect is similar to producing a report. To sort by more than one column, you must first arrange and then select the columns that you'll use for sorting; the columns must be arranged in order from left to right. The values in the selected column that's farthest left will be used to sort the records first, and the values in the selected column that's farthest right will be used to sort the records last. The columns for sorting aren't required to be adjacent to each other; however, if they are, the order of the records is more obvious.

Sorting records by one column

1. Click the heading of the column whose values you want to use to sort the records.

2. Right-click the selected column's heading and click Sort Ascending or Sort Descending.

   The table's records are sorted.

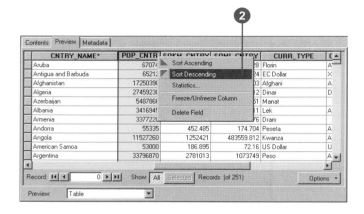

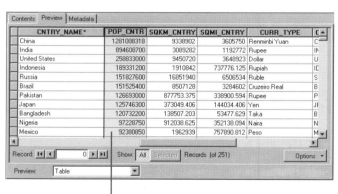

The records have been sorted according to the selected column's values.

Sorting records by more than one column

1. Rearrange the table's columns so that the column whose values will be used to sort the records first appears to the left of the column whose values will be used second, and so on.

2. Click the heading of the first column you want to use to sort the records.

3. Press the Ctrl key on the keyboard and click the second column's heading.

4. Repeat step 3 until all columns that will be used to sort the table's records have been selected.

5. Right-click the heading of one of the selected columns and click Sort Ascending or Sort Descending.

    The table's records are sorted.

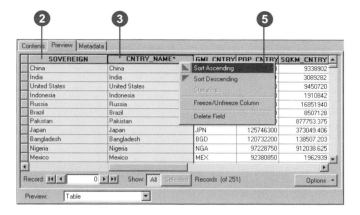

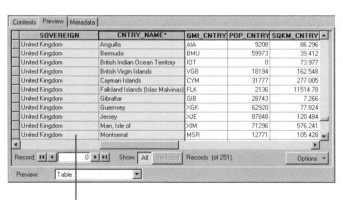

The records have been sorted first by the left column's values and then by the right column's values.

Finding values in a table

Occasionally, you have to look for records in a table that contain specific text or numbers. By default, the Find command tries to match your text to any part of the table's values. The text "San" would be found in "Pleasants" and in "San Juan". However, if you choose Start of Field from the Text Match dropdown list in the Find dialog box, "Pleasants" will not be found. To find all instances of the exact text "San Juan", choose Whole Field instead. You can search for numbers the same way you search for text.

When looking for a record, Find assumes by default that the current record is at the top of the table and that you want to search down through the remaining records. If the current record is at the bottom of the table, you may want to search up through the table's records instead—click Up in the Search dropdown list. If you choose All, the search starts from the current record, goes down through the remaining records in the table, loops around to the top of the table, and then proceeds down to the current record.

Finding text in a column

1. Click the heading of the column that contains the text for which you want to search.

2. Click Options and click Find.

3. Type the text you want to find in the Find what text box.

4. Click Find Next.

   The first record found containing your text is selected.

5. If you want to find another record containing the same text, click Find Next again.

6. Repeat step 5 until you are finished looking for values in the table.

7. Click Cancel.

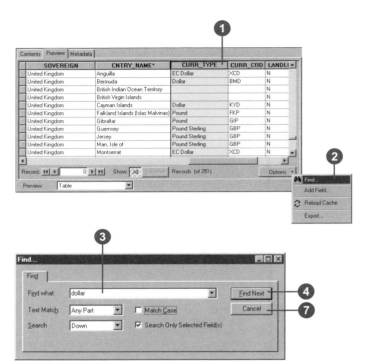

The first record found containing your text is selected.

Matching the case

To match the capitalization of the text you type, check Match Case in the Find dialog box; if you don't want to match the text's capitalization, uncheck Match Case. The Match Case option won't be available for some databases.

Finding text in any column in the table

1. Click Options and click Find.

2. Type the text you are looking for in the Find what text box.

3. Uncheck Search Only Selected Field(s).

4. Click Find Next.

   The first record found which contains your text is selected.

5. If you want to find another record containing the same text, click Find Next again.

6. Repeat step 5 until you are finished looking for values in the table.

7. Click Cancel.

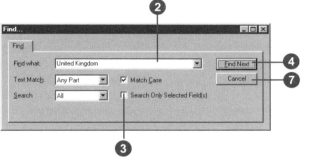

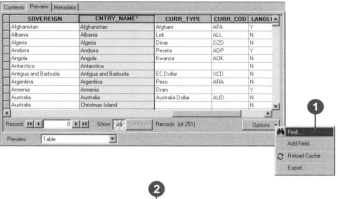

The first record found containing your text is selected.

Adding and deleting columns

When exploring a table's contents, you may decide that a column should be deleted or that a new column should be added. Table view provides an easy way to do this without having to open the item's Properties dialog box. This technique for adding and deleting columns works exactly the same for all data sources regardless of their format. You must have write access to the data to add or delete a column.

Adding a column to a table

1. Click Options and click Add Field.

2. Type the name of the new column.

3. Click the Type dropdown arrow and click the appropriate data type for the new column.

    The properties that are appropriate to the new column's data type appear in the Field Properties list below.

4. Set the properties for the new column. For example, type the maximum number of characters that Text values can have.

5. Click OK.

    The new column appears to the right of all other columns in the table.

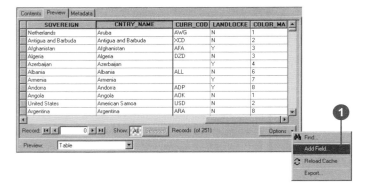

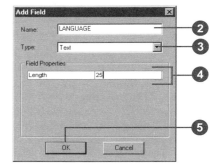

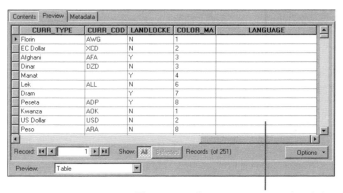

The new column appears to the right of all other columns in the table.

Deleting a column from a table

1. Click the heading of the column that you want to delete.

2. Right-click the selected column's heading and click Delete Field.

   A warning message appears indicating that deleting columns cannot be undone.

3. Click Yes to delete the column.

   The column is removed from the table.

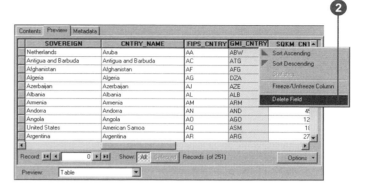

The column is removed from the table.

Creating new data sources from the values in a table

You may decide when looking at a table to create a copy of its records. Table view lets you export the data source's attributes and in doing so create a new table. Regardless of the type of data source you're looking at, you can create a new dBASE, INFO, or geodatabase table.

Similarly, if a table contains x,y,z coordinates, ArcCatalog lets you create point features representing those locations quickly and easily. The Create Feature Class From XY Table dialog box lets you pick which columns contain the coordinate values, define their spatial reference, and choose the format in which you want to create the new point locations.

Exporting records

1. Click Options and click Export.

2. In the Export Data dialog box, click the Browse button.

3. Click the Save as type dropdown arrow and click the format to which you want to export the data. For example, click Personal Geodatabase tables.

4. Navigate to the folder or geodatabase in which you want to place the exported data.

5. Type a name for the new data source.

6. Click Save.

7. Click OK.

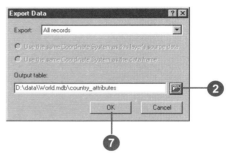

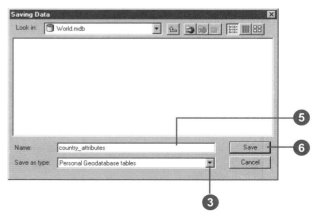

Creating point features from coordinate values

1. Right-click a table in the Catalog tree that has columns containing x,y coordinates. The table may also have a column containing z-coordinates.

2. Point to Create Feature Class and click From XY Table.

3. Click the X Field dropdown arrow and click the name of the column that contains the x-coordinates.

4. Click the Y Field dropdown arrow and click the name of the column that contains the y-coordinates.

5. If appropriate, click the Z Field dropdown arrow and click the name of the column that contains the z-coordinates.

6. Click Spatial Reference of Input Coordinates to define the coordinate system for the input values.

7. Follow the steps for 'Defining a shapefile's coordinate system' in Chapter 12 to define the spatial reference of the coordinates.

8. Click the Browse button. ▶

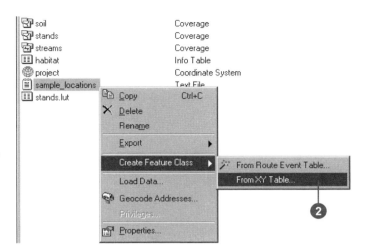

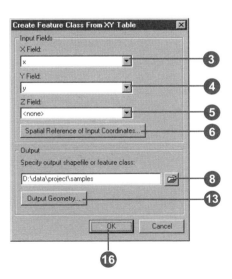

9. Click the Save as type dropdown arrow and click the format in which to create the new point features. For example, click Shapefile.

10. Navigate to the folder or geodatabase in which you want to store the new features.

11. Type a name for the new data source.

12. Click Save.

13. If the output features should have a different coordinate system from the input values, click Output Geometry.

    Otherwise, skip to step 16.

14. Click the ellipses (...) button to the right of the Spatial Reference property and define the spatial reference of the output features.

15. Click OK, then click OK in the Define Output Geometry dialog box.

16. Click OK.

    A new point data source is created in the appropriate location.

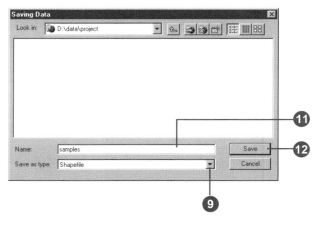

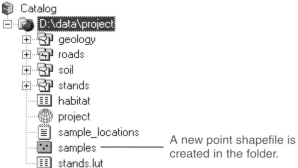

A new point shapefile is created in the folder.

Working with metadata

9

Metadata is information that describes your data in the same way a card in a library's card catalog describes a book. After you create new data, you should create metadata to document its contents. When detailed metadata has been created, it can answer your questions and help you make decisions. For example, it can help you determine when your data is out of date, what map scale is appropriate for presenting your data, or how accurate your data is—and therefore how much confidence you can have in your conclusions.

Any item in the Catalog, including folders and file types such as Word documents, can have metadata. With the Catalog's default settings, all you have to do to create metadata is click the item in the Catalog tree and click the Metadata tab—the Catalog both creates metadata and records the item's properties within it. Once created, metadata is copied, moved, and deleted along with the data source when it is managed with ArcCatalog or ArcInfo Workstation. The first two tasks in this chapter are for anyone who wants to explore metadata. The rest of the chapter is helpful if it is a user's job to document the data.

Exploring an item's metadata

To decide if a data source is suitable, you often need more information than is available in Details view or the Properties dialog box. You may need to know how accurate the data is or how a set of measurements was collected. You'll find descriptive information like this in the Metadata tab.

Click the tabs in the metadata to see different categories of information. The Description tab includes information about the status of the data source, its location, and any enclosed files. The Spatial tab shows the data's extent, as well as detailed feature or raster properties. The Attributes tab describes each attribute and lists the relationships in which the data source participates. Within the tabs, click a green heading to hide or show its information. For example, in the Attributes tab, click an attribute name to see a description of its values and its data type. Click the attribute name again to hide those details. When you position the mouse pointer over a heading, the text color changes and the pointer changes to a hand.

1. Click the item in the Catalog tree for which you want to see metadata.

2. Click the Metadata tab.

3. Click a tab on the metadata page to see a different group of metadata elements.

4. Click a group heading to hide its contents.

   Click the heading again to show its contents.

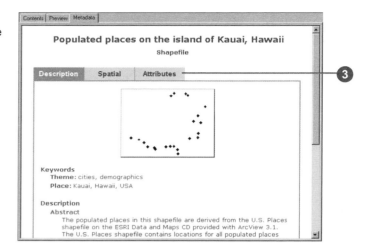

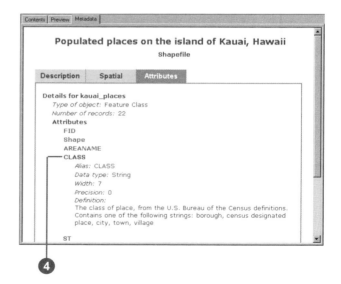

Changing the metadata's appearance

A stylesheet is similar to a query, which selects data from a database. Stylesheets select which metadata elements to display and define how their values appear. Each stylesheet in the Catalog presents the same body of metadata using a different set of rules. Four stylesheets are provided: ESRI, FGDC, FGDC FAQ, and XML. You can add your own custom stylesheets to the list. The metadata's default appearance is defined by the ESRI stylesheet. You can change its appearance by choosing a different stylesheet in the dropdown list on the Metadata toolbar.

The FGDC stylesheet shows all metadata elements defined by the CSDGM. Authored by the U.S. FGDC, this book refers to it as the FGDC standard. The FGDC stylesheet's format will be familiar to you if you've worked with the FGDC standard or searched for data using the National Spatial Data Institute (NSDI) Geospatial Data Clearinghouse. The FGDC FAQ stylesheet presents a subset of the FGDC metadata elements in question and answer format; it may be helpful if you're new to ▶

Changing the current stylesheet

1. Click the Stylesheet dropdown arrow on the Metadata toolbar and click a different stylesheet.

   The metadata's appearance changes according to the rules of the new stylesheet.

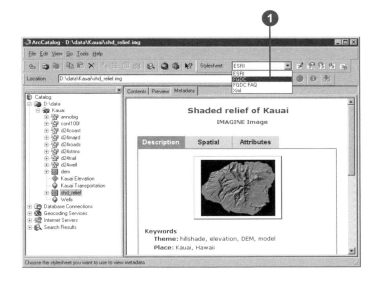

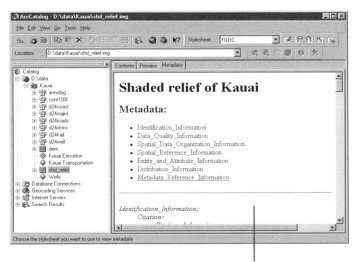

The metadata's appearance changes according to the new stylesheet's rules.

metadata. These stylesheets emulate the HTML pages that you can create with the FGDC's metadata parser utility, mp. The ESRI stylesheet shows many elements defined in the FGDC standard, in addition to detailed properties of the data which are defined by ESRI.

The XML stylesheet shows the entire contents of an item's metadata (or any other XML document). Different colors let you easily distinguish between an element's name and its value. XML data is hierarchical by nature; instead of a value, an element may contain other elements. The XML stylesheet shows plus or minus signs next to group elements; you can click them to hide or show the elements they contain.

Every time you start ArcCatalog, metadata is initially presented with the default stylesheet. To change the default stylesheet, use the Options dialog box. The next time you start ArcCatalog, metadata will initially be presented with the stylesheet of your choice. Setting the default stylesheet doesn't prevent you from changing the current stylesheet using the dropdown list on the Metadata toolbar.

Setting the default stylesheet

1. Click the Tools menu and click Options.

2. Click the Metadata tab.

3. Click the Default Stylesheet dropdown arrow and click the stylesheet that should be the default.

4. Click OK.

   The next time you start ArcCatalog, the stylesheet you've chosen will be used as the default.

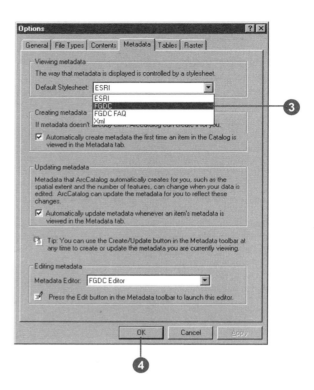

Metadata in ArcCatalog

Metadata in ArcCatalog consists of properties and documentation. *Properties,* such as the extent of a shapefile's features, are derived from the item itself. *Documentation* is descriptive information supplied by a person.

By default, when you attempt to view an item's metadata, ArcCatalog will create it automatically if metadata doesn't already exist; it will then add many of the item's properties to it. Hints about the documentation you should provide are also added. Once created, metadata becomes part of the item itself. It is automatically moved, copied, and deleted along with the item.

Every time you view the metadata, the Catalog automatically updates the properties recorded in it with current values. This ensures that the metadata is kept up to date with changes to the data source. For example, the extent and count of a shapefile's features will be current when you look at its metadata, even if new features were recently added.

If you want more control over when metadata is created and updated, you have a few choices. You can turn off automatic updates for individual items—for example, after they have been published. Then, you can choose to create and update metadata manually by clicking the Create/Update Metadata button on the Metadata toolbar.

Writing documentation

Documenting your data protects your organization's investment in that data. Without knowledge of the data's accuracy, provenance, and age, you can't have a high level of confidence in decisions based on that data. Creating detailed metadata describing these qualities ensures that you can continue to use your data and make decisions based on it.

In its simplest form, documentation might be a few lines in a text file. While better than nothing, this type of documentation may only be useful to the person who created it. For example, steps in the processing history may be omitted because they seemed obvious at the time. It is important that other people clearly

understand your documentation—particularly information describing how your data should and should not be used.

Because errors of omission and misinterpretation can be costly, efforts have been made to create standards for documenting spatial data. Standards range from simple to complex. Different states and countries have created their own standards to try to simplify and clarify the information that should be recorded. However, the proliferation of standards has caused confusion.

Fortunately, the *International Standards Organization (ISO)* is creating a unified content standard. With a common standard it will be easier for those within an organization, as well as the GIS community as a whole, to create and use metadata.

Simply because so many different standards exist, metadata in ArcCatalog isn't required to meet any specific one. However, standards can be enforced by a metadata editor. The metadata editor provided with the Catalog lets you document your data following the *FGDC* standard. If you complete the documentation suggested by the Catalog's hints, your metadata will satisfy the FGDC standard's minimum requirements.

The ESRI profile of the FGDC standard

The FGDC standard attempts to describe all types of geographic data in general terms. ESRI has defined many properties of items that aren't included in the FGDC standard such as properties of geometric networks and relationship classes. These elements belong to the *ESRI Profile of the Content Standard for Digital Geospatial Metadata.* Documentation about the ESRI profile is available at *www.esri.com/metadata.*

When ArcCatalog creates and updates metadata, the item's properties and documentation hints are added to the appropriate elements as defined by the FGDC standard and the ESRI profile. Your metadata will be compliant with the FGDC standards even with the additional elements defined by the ESRI profile because the Catalog notes in the item's metadata that the ESRI profile is being used. ▶

You can create custom editors that let you document your data following different standards or add custom content for your organization. For example, a company could add an element recording business procedures such as whether or not a data source has passed a quality control review. If you build an editor that adds custom elements to the metadata and you want to see them in ArcCatalog, you must also build new, or customize existing, stylesheets to include these elements.

How metadata is stored

Metadata created with ArcCatalog is stored as XML data either in a file alongside the item or within its geodatabase. XML is a markup language similar to HTML. HTML defines both the data and how it's presented. XML, on the other hand, lets you define data using tags that add meaning.

Stylesheets define how XML data is presented. They are created using XSL. XSL elements query and evaluate XML data. A stylesheet retrieves specific values from an XML document, formats them, and then defines how they are presented. ArcCatalog stylesheets generate HTML pages from XML data. Therefore, you can interact with metadata as you would interact with any HTML page in a browser.

Because the presentation information is stored separately, you can change the metadata's appearance using different stylesheets. To change the stylesheet that is currently being used to transform the metadata, simply click a different stylesheet from the dropdown list in the Metadata toolbar.

Each metadata element defined in the FGDC standard or the ESRI profile has both a long and a short name. The short name is used as the name of the metadata's associated XML element. For example, the Cloud Cover element in the FGDC standard has the the short name cloud; its associated XML element is <cloud>.

XML elements can have attributes. ArcCatalog refers to the Sync attribute before it updates an element's value. When the Catalog initially records an item's properties in its metadata, the Sync attribute for the associated element is set to "TRUE", for ex-

ample, <title Sync="TRUE">. When the Catalog updates metadata, if it doesn't find the Sync attribute, or its value isn't "TRUE", it won't overwrite the element's value. If you modify the item's properties with the metadata editor, the Sync attribute is removed; the value you've typed won't be overwritten. You can also modify an element's XML attributes programmatically; this gives more control over how and when metadata is updated.

XML is an emerging industry standard, that is being used to transfer data across the Internet. For example, it is often used in e-commerce transactions. Many different applications can be used to view or edit metadata XML files.

Metadata for folders and geodatabases

You can document the contents of a folder or geodatabase using a metadata editor in the same way you would create documentation for any other item in the Catalog. You might describe the project for which the data was created, describe the general area in which the data is located, or give a warning to users that the data in the folder is incomplete. This type of information can be helpful when people are browsing for data.

If you prefer, you can create an HTML page to document a folder's or geodatabase's contents. You might choose a colorful graphic design that targets people in your organization who are not in your GIS group. It can help them locate the maps and layers they need on the network.

For folders, place a "metadata.htm" file within the folder it describes. For geodatabases, create an .htm file and then import its contents using the XML importer. Like XML, these HTML pages must be well-formed and free of syntax errors. To be well-formed, all opening tags such as <P> for paragraph must be properly closed by an end tag such as </P>. Unfortunately, most HTML authoring tools don't create well-formed HTML. You may have to fix your .htm files in a text editor before they will display properly in the Catalog.

Defining when metadata is created and updated

By default, metadata is automatically created or updated when you view it in the Metadata tab. However, you can choose to create and update all metadata manually. When you change these settings in the Options tab, it affects how metadata is handled for all items in the Catalog. You can turn off automatic updates for specific items using the Metadata Properties dialog box. You might do this after completing and publishing their metadata so it is not accidentally changed later.

You can manually update an item's metadata at any time by clicking the Create/Update Metadata button on the Metadata toolbar. Suppose that while you are viewing an item's metadata, you modify its properties. Click the Create/Update Metadata button to have those changes reflected in the metadata.

Choosing how metadata is created and updated

1. Click the Tools menu and click Options.

2. Click the Metadata tab.

3. Check the appropriate boxes to have the Catalog automatically create or update metadata.

   Uncheck the appropriate boxes if you prefer to manually create or update metadata.

4. Click OK.

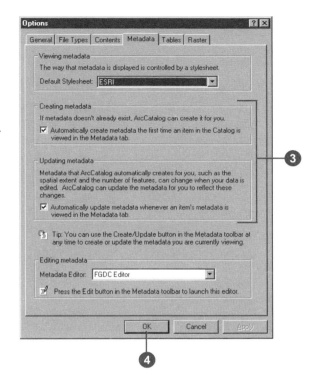

Turning off automatic updates for individual items

1. In the Catalog tree, click the item for which metadata should not be updated automatically.

2. Click the Metadata tab.

3. Click the Metadata Properties button on the Metadata toolbar.

4. Click the Options tab.

5. Check Do not automatically update metadata.

6. Click OK.

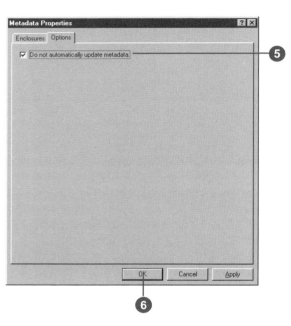

Creating and updating metadata manually

1. In the Catalog tree, click the item whose metadata you want to create or update.

2. Click the Metadata tab.

3. Click the Create/Update Metadata button on the Metadata toolbar.

Editing metadata

The Catalog comes with an FGDC metadata editor. To start using the FGDC metadata editor, click the Edit Metadata button on the Metadata toolbar. If more than one editor is available—for example, if you've created a custom editor—you can choose to have a different metadata editor appear when you click the Edit Metadata button. The available metadata editors are listed in the Options dialog box. Although you can only use one at a time, you can use different editors to document your data. Because metadata for coverages, shapefiles, and other file-based data sources is stored as XML files on disk, you can also use XML editors or Visual Basic® (VB) applications to edit their contents outside of ArcCatalog. For example, you might do this to create a template of information that can be imported into many different data sources such as how to purchase the data or who to contact for more information.

Adding documentation

1. Click the item whose metadata you want to edit in the Catalog tree.

2. Click the Metadata tab.

3. Click the Edit Metadata button on the Metadata toolbar.

4. Document your data using the metadata editor.

5. Close the metadata editor.

Choosing a metadata editor

1. Click the Tools menu and click Options.

2. Click the Metadata tab.

3. Click the Metadata Editor dropdown arrow and click the editor you want to use.

4. Click OK.

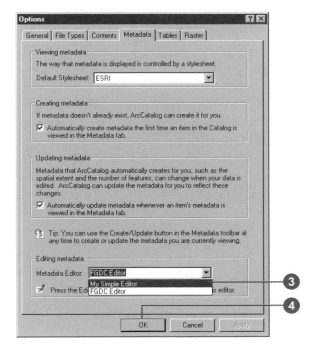

Using the FGDC metadata editor

The FGDC metadata editor lets you create FGDC-compliant metadata for the selected item in the Catalog tree. This is the default metadata editor in ArcCatalog.

The information in an FGDC metadata document is divided into seven main sections: Identification, Data Quality, Spatial Data Organization, Spatial Reference, Entity and Attribute, Distribution, and Metadata Reference. The section titles across the top of the FGDC metadata editor correspond to these sections. Information within each section is divided into subsections. When you click a section title, several tabs appear, with each tab representing a subsection.

You can enter values for all elements defined in the FGDC standard and some elements defined in the ESRI profile using this editor. At first, the organization of elements may seem confusing,

but as you become familiar with the FGDC standard you will find it easier to locate the elements you want to edit.

For most elements, text boxes are provided in which you can type the appropriate information. When the standard expects one of several predefined values, a dropdown list lets you choose one of the appropriate values. For some elements, you can either choose a predefined value or type your own information. For example, to define the time period of the data, choose "unknown" from the list or type an appropriate date.

Some properties are added to the metadata by ArcCatalog. These are best left unmodified and are read-only in the editor. However, you can edit other properties. Suppose many features are grouped together and a few are set apart. The bounding coordinates added by the Catalog represent the extent of all features. You might edit the coordinates to represent the area occupied by the majority of the features. After editing elements that were added by the Catalog, they won't be updated automatically in the future.

Some elements can be repeated many times. For example, you might use several keywords to describe the data's subject. There is a toolbar containing buttons that let you add, delete, and browse through your entries. For each repeating element the toolbar's status line shows the name of the repeating element and indicates both the total number of entries for that element and which entry is currently displayed. The appropriate navigation buttons are unavailable when you're at the first or last entry. ▶

Click a section title to edit the elements in that section. The selected Section Title has bold text.

Click a tab to edit the elements in that subsection.

Type information for a metadata element or choose a value from a dropdown list.

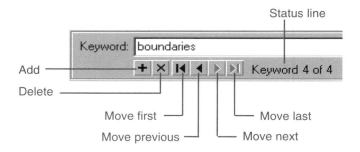

Status line

Add

Delete

Move first

Move previous

Move last

Move next

When a single element can be repeated, the toolbar is located immediately below the text box. When a group of elements can be repeated, the toolbar is located at the bottom of the group. For example, a shapefile can have many attributes, and several elements are used to describe each one; a toolbar controls the group of elements that are repeated for each attribute. Similarly, coverages have many feature classes. A different toolbar controls the group of elements describing the feature class, which includes the group of elements describing its attributes.

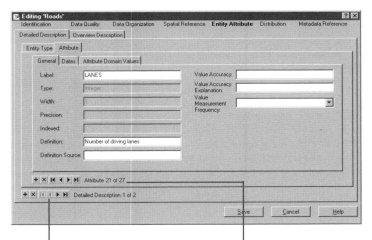

The Entity Type tab contains a group of elements describing the current feature class. With the toolbar, you can manage metadata for each feature class in a coverage.

This toolbar lets you manage metadata for each attribute in the current feature class. All elements on the Attribute tab describe the current attribute.

Getting Help in the editor

For a detailed description of a section title, or any metadata element, click the What's This? button in the upper-right corner, then click the section title or element in the dialog box. A Help description box appears showing the type of information that is expected, along with an indication of whether it is considered mandatory, mandatory-if-applicable, or optional in the FGDC standard.

Clicking Help in the editor shows you a topic that walks you through the process of completing the mandatory information in the FGDC standard. Derived from the *Content Standard for Digital Geospatial Metadata Workbook,* it shows you where to find the appropriate elements in the metadata editor and describes the information you should provide.

Use the workbook itself to get a better understanding of the FGDC standard and information about how to create a complete FGDC metadata record. It is available for download from the NSDI Web site at *www.fgdc.gov/metadata*. A link to this Internet site is provided from the editor's Help topic. All U.S. government agencies and state and local agencies that receive federal funds to create metadata are mandated to follow the FGDC standard.

Adding enclosures

Documentation for an item, such as a Word document describing how values were collected, may already exist. You can enclose a copy of a file within an item's metadata; in most cases, this should be a preliminary measure until you have time to create detailed metadata using the editor. If the original file is later modified, the copy stored within the metadata won't change. Enclosing files in an item's metadata works the same way as enclosing files in an e-mail message.

You might enclose a layer file that can be added to a map document. You might also enclose an image in a folder's metadata to describe the general area in which its data is located. When enclosing an image, check Image in the Add New Enclosure dialog box. Enclosed files are listed in the ESRI stylesheet. For enclosed images, the ESRI stylesheet will also show thumbnails describing their contents. If metadata doesn't already exist when you add an enclosure, metadata will be created.

Enclosing documentation files

1. In the Catalog tree, click the item to which you want to attach a file.

2. Click the Metadata tab.

3. Click the Metadata Properties button on the Metadata toolbar.

4. Click the Enclosures tab.

5. Click the Add button.

6. Click the Browse button.

7. Navigate to and click the file you want to attach. Click OK.

8. Type a description of the file.

9. If the enclosed file is an image, such as a Windows Bitmap file, check Image.

10. Click OK. ▶

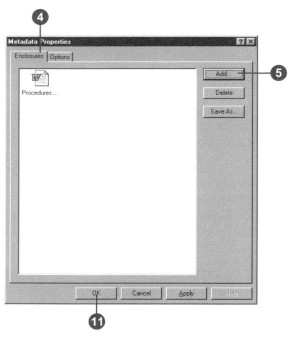

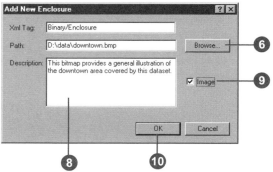

11. Click OK.

Enclosed files are listed in the ESRI stylesheet. You can also see thumbnails of enclosed files that are images.

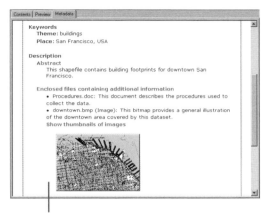

The ESRI stylesheet lists enclosed files and shows thumbnails of image enclosures.

Viewing an enclosed file

1. Click the Metadata Properties button on the Metadata toolbar.

2. Click the Enclosures tab.

3. Double-click the enclosed file whose contents you want to see.

   The file opens in the appropriate application.

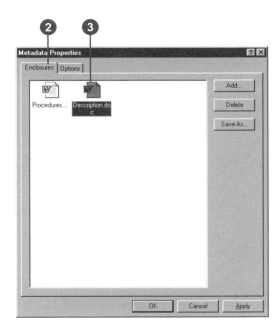

Importing and exporting metadata

While you can create metadata in ArcCatalog, you may already have metadata that was created with another metadata tool. If it's stored in the text, XML, or SGML format defined by the FGDC's metadata parser utility, mp, you can import that metadata. ArcCatalog uses mp itself to import and export metadata in its supported formats; these are the FGDC CSDGM formats in the Format list. The XML format can be used to import metadata that was created with ArcCatalog. Importing overwrites all of an item's existing metadata.

Suppose you import metadata created with another application. By default, ArcCatalog updates as many metadata elements as possible with current values derived from the item such as the extent and coordinate system of its data. These values will be updated automatically every time you view the metadata. However, if you uncheck Enable automatic update of metadata, ArcCatalog won't overwrite any existing values. If you want to choose specific elements that should be updated by the Catalog, use ▶

Importing metadata

1. In the Catalog tree, click the item for which you want to import metadata.

2. Click the Metadata tab.

3. Click the Import Metadata button on the Metadata toolbar.

4. Click the Format dropdown arrow and click the format of the metadata that you will be importing.

5. Click the Browse button.

6. Navigate to and click the metadata file whose contents you want to import. Click Open.

7. If you don't want ArcCatalog to update the metadata with the current properties of the data, uncheck Enable automatic update of metadata.

8. Click OK.

   The imported metadata appears in the Metadata tab.

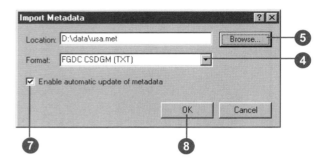

The imported metadata appears in the Metadata tab.

the metadata developer sample Set Synchronized Properties or set the Sync attribute for the appropriate elements to "TRUE" programmatically.

You might export metadata that is maintained by ArcCatalog to publish it on the NSDI's geospatial data clearinghouse. When exported to the FGDC CSDGM formats, the elements defined in the ESRI profile are excluded. Because ArcCatalog uses mp to create the resulting files, they can be posted directly on the clearinghouse. Exporting to the format "HTML" creates a file that represents the selected item's metadata exactly as you see it in the Metadata tab. The format "XML" creates a copy of the item's metadata in a new XML file. This lets you work with metadata for geodatabase items outside ArcCatalog.

If your organization has defined its own metadata standard or a metadata database, you can create custom importers and exporters that translate between the Catalog's XML and your custom format. You may want to do this if you want the Catalog to automatically maintain the metadata, but other applications require metadata in your format. Once registered with the Catalog, your importer or exporter will appear in the list in the Import or Export dialog box.

Exporting metadata

1. In the Catalog tree, click the item for which you want to export metadata.

2. Click the Metadata tab.

3. Click the Export Metadata button on the Metadata toolbar.

4. Click the Browse button.

5. Navigate to the folder in which the exported metadata should reside, type a name for the new metadata file, then click Save.

6. Click the Format dropdown arrow and click the format in which you want to export the metadata.

7. Click OK.

A new file containing a copy of the item's metadata is created in the appropriate format.

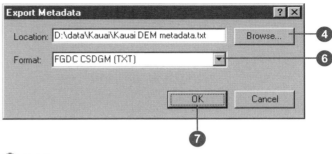

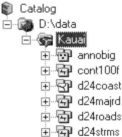

A new file is created containing a copy of the item's metadata.

Searching for items

10

Sometimes you know where data is located; other times you don't. It can be time-consuming to browse through each shared folder on the network looking for the data you need. The Catalog's Search tool can solve that problem.

You can search for any item that you can use in the Catalog by its name and the type of data it contains. You can search for geographic data that is located within a specific area on the earth's surface. If the item has metadata, you can search by temporal information such as when the data was published or by keywords such as the name of the person who created the data. The Catalog creates shortcuts to the items that have been found and places them in the search results list.

Searching for items

When searching for items in ArcCatalog, you can specify the name and type of item you want to find. When defining criteria that will test an item's name, you can use an asterisk (*) to represent one or more characters. For example, "San*" would return all items whose name begins with "San" including San Francisco and Santa Fe. You can also choose the specific types of items you want to find in the Type list.

When defining where to start looking for items, you can choose to search in the Catalog, the default, or in the File system. When searching the Catalog you can look in any folder, database connection, or Internet server. When searching the Catalog, if an item doesn't have metadata, you can only check its name, type, and geographic location. If metadata is present, you can also search using temporal and keyword criteria, and the name, type, and location information is derived from the metadata rather than the item itself.

When searching the File system, your search may be faster than searching the Catalog, but you will only find items for which metadata ▶

Defining the name, type, and location of the items you want to find

1. Click the Search button on the Standard toolbar.

2. Click the Name & location tab.

3. If you want to search by name, type all or part of the name you're looking for into the Name text box. Use an asterisk (*) to represent one or more letters. Otherwise, type an asterisk in the Name text box.

4. If you want to search by type, press the Ctrl key while clicking the items you want to find in the Type list. Otherwise, click Clear.

5. Click the Search dropdown arrow and click where you want to search.

6. Click the Browse button.

7. Navigate to and click the folder, database connection, or Internet server in which to start searching. Click Open.

8. Type a name for your search in the Save as text box.

9. Click Find Now. ▶

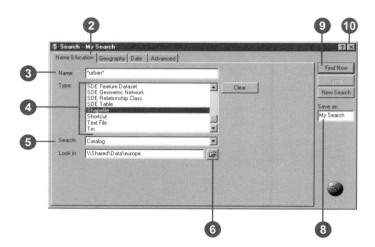

exists. With a File system search, you can't search the contents of databases or Internet servers.

When you click Find Now in the Search dialog box, your search is saved in the Search Results folder and is automatically selected in the Catalog tree. When an item is found that satisfies your search criteria, a shortcut to that item is added to the search results list. You can close the Search dialog box during the search. If your criteria were too inclusive, the Catalog may find a very large number of items; you can stop the search, redefine your criteria, and then start searching again.

The Search is saved in the Search Results folder and is selected in the Catalog tree. As items are found that satisfy the search criteria, shortcuts to those items are added to the search results list.

10. Click the Close button in the upper-right corner of the Search dialog box.

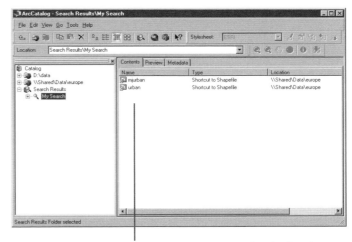

As items are found that satisfy your search criteria, shortcuts to those items are added to the search results list.

Stopping an ongoing search

1. Click the Search button on the Standard toolbar or right-click the ongoing search and click Properties.

   The Search dialog box reappears.

2. Click Stop.

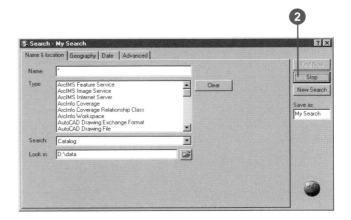

Searching with geographic criteria

In addition to searching by name and type, you can search for data that covers a specific geographic area. You can draw your area of interest on a map or choose a place name from the dropdown list. You can use different maps from the Map dropdown list to define your geographic criteria. If the scale of the available maps is too small or if they don't cover the appropriate area, click "<Other...>" in the Map dropdown list and then select a data source of your own to use. For example, you might use a satellite image of your city to define your geographic criteria.

Defining your area of interest

1. Click the Search button on the Standard toolbar.

2. Click the Name & location tab to define the name, type, and location criteria for your search.

3. Click the Geography tab in the Search dialog box.

4. Check Use geographic location in search.

5. Draw a box on the map outlining your area of interest.

   Or click a place name in the Choose a location dropdown list. A box will be drawn on the map around that location.

   Or type the North, South, East, and West coordinates that define your area of interest. The box on the map changes accordingly.

6. If you wish, modify the box using the Selection tool or remove the box and start again.

7. Click Find data entirely within location or Find data overlapping location—whichever is appropriate for your search.

8. Type a name for your search in the Save as text box.

9. Click Find Now.

10. Click the Close button.

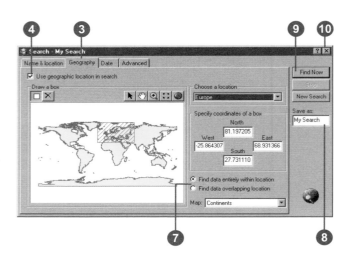

Using your own data as a map

1. Click the Map dropdown arrow and click "<Other...>".

2. Navigate to and click the appropriate data source. Click Add.

   The data source is added to and selected in the Map dropdown list. The geographic data is drawn in the map, and a box that encloses the data appears.

3. Modify the box so that it contains your area of interest.

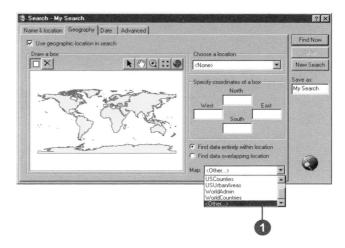

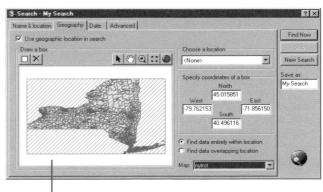

Your data source appears in the map along with a box that encloses the data.

Searching with temporal criteria

The Catalog lets you search for items using temporal criteria. You can search for items according to how current the data is, when the item's metadata was last updated, or when the item was published. Temporal information about an item is derived from its metadata. Searches can either compare dates in the metadata to a specific date, a range of dates, or a general period of time. To use a specific date, click Before, Equal to, or After in the dropdown list and define the date; the Catalog will ignore the ending date in the date range. When using a general period of time—for example, the previous 60 days—the time period is always calculated in relation to today's date. In the Search dialog box, dates appear according to the fomat of the system date on your computer.

Defining a specific date

1. Click the Search button on the Standard toolbar.

2. Click the Name & location tab to define the name, type, and location criteria for your search.

3. Click the Date tab in the Search dialog box.

4. Click Find and click the Find dropdown arrow. Click the type of date on which you want to search.

5. Click the dropdown arrow and click Before, Equal to, or After.

6. To use a date other than today, click the first date dropdown arrow. Navigate to and click the appropriate date.

   Or click the month, day, or year and type a new value. The new date appears in both date text boxes. When specifying the year, you can either type a two- or four-digit date.

7. Type a name for your search in the Save as text box.

8. Click Find Now.

9. Click the Close button to close the Search dialog box.

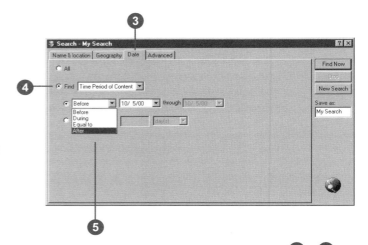

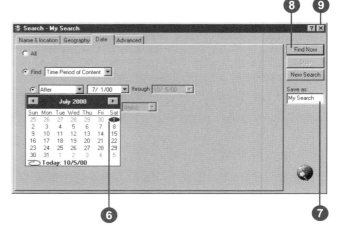

Defining a range of dates

1. Click the Search button on the Standard toolbar.

2. Click the Name & location tab to define the name, type, and location criteria for your search.

3. Click the Date tab in the Search dialog box.

4. Click Find and click the Find dropdown arrow. Click the type of date for which you want to search.

5. Click the dropdown arrow and click During.

6. Define the beginning date.

7. Define the ending date.

8. Type a name for your search in the Save as text box.

9. Click Find Now.

10. Click the Close button to close the Search dialog box.

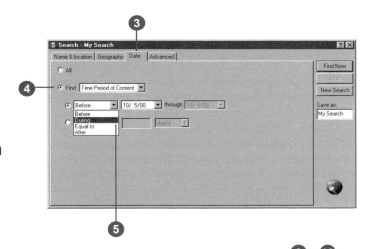

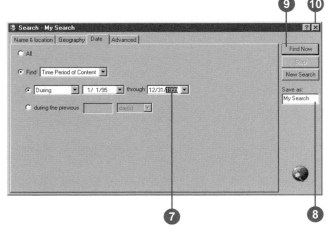

Defining a general range of time

1. Click the Search button on the Standard toolbar.

2. Click the Name & location tab to define the name, type, and location criteria for your search.

3. Click the Date tab in the Search dialog box.

4. Click Find and click the Find dropdown arrow. Click the type of date for which you want to search.

5. Click during the previous.

6. Click the dropdown arrow and click the appropriate period of time.

7. Type the appropriate duration into the text box.

8. Type a name for your search in the Save as text box.

9. Click Find Now.

10. Click the Close button to close the Search dialog box.

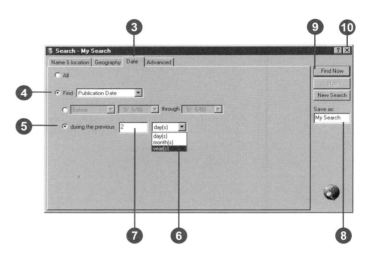

Searching with keywords

You can search for items whose metadata contains specific text. You might look for items that have a specific word in their Title or in the Abstract that describes the item's contents. With Full Text, the Catalog will locate items where a word exists anywhere in their metadata. When you define several keyword criteria, they are combined using the Boolean operator AND; for example, look for items whose Abstract includes the word "soil" AND the word "salinity".

Several metadata elements that are commonly used in searches appear in the Metadata element list in the Advanced tab. If you are interested in information stored in a metadata element that isn't in this list, you can type the element's "path" in the Metadata element text box. An element's path, like a file's path, describes how to navigate the hierarchy of the metadata XML document from its root to find the element. For information on how to find a metadata element's path, see Chapter 6, 'Managing the Catalog's contents'.

1. Click the Search button.

2. Click the Name & location tab to define the name, type, and location criteria for your search.

3. Click the Advanced tab in the Search dialog box.

4. Click the Metadata element dropdown arrow and click the appropriate keyword to use from the metadata.

5. Click the Condition dropdown arrow and click the appropriate condition.

6. Click in the Value text box and then type the words for which you want to search. Or click the Value dropdown arrow and click the appropriate Value in the list.

7. Click Add to List.

8. Repeat steps 4 through 7 to add additional criteria to the list.

9. If you don't want to search on a criterion, click it and click Delete to remove it from the list.

10. Check Match case to return items whose metadata contains the values exactly as they have been typed.

11. Type a name for your search in the Save as text box.

12. Click Find Now.

13. Click the Close button.

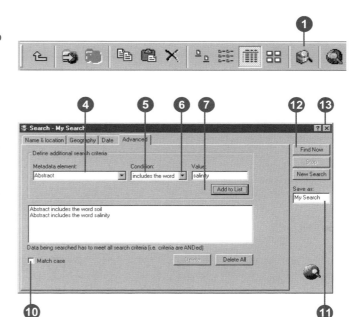

Exploring the results of your search

You can explore search results in the Catalog the same way that you explore the contents of other items. When a Search is selected in the Catalog tree, you can list its shortcuts in the Contents tab. In the Metadata tab, you'll see a detailed description of the criteria that were defined for the search.

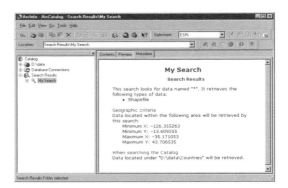

In the Preview tab, you'll see an overview describing where on the earth's surface the shortcuts' data is located. Use the Identify tool to determine which data sources are located in which areas, then select the ones that are of interest to you.

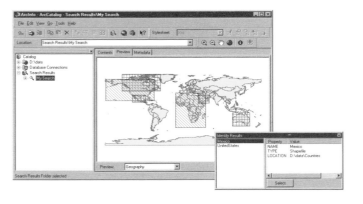

A search result is comprised of the shortcuts that were found by the Catalog. These shortcuts, like Windows shortcuts, provide a link to an item that is stored elsewhere. When a shortcut is selected in the Catalog tree, you can preview the item's data and metadata in the appropriate tabs as if you were exploring the contents of the item itself.

Once you've found the item that you want to use, you can right-click the shortcut and click Go To Target to select the item itself in the Catalog tree. Alternatively, you can work with the shortcut as if you were working with the item itself. You can drag and drop a shortcut onto a map or an ArcToolbox tool. You can also double-click a shortcut to open the item's Properties dialog box. And, if you have write access to the data, you can modify its properties and update its metadata.

Modifying the search results

One of the reasons why you might be searching for items is so you can see which ones must be updated. For example, after repairing the contact information in the metadata for several items, you might update the search results so that you have an accurate list of your outstanding work. You can either delete each shortcut from the search results list as you modify its metadata, or you can rerun the search. Shortcuts in ArcCatalog are like Windows shortcuts—deleting, copying, and moving the shortcut has no influence on the original item. To copy the original item, right-click it and click Copy Target, then click the destination and click Paste. To delete the original item, select it in the Catalog tree and then click Delete; remember to delete the shortcut as well.

Updating the search results

1. Right-click a Search and click Properties.

   The Search dialog box appears containing the Search's criteria.

2. Modify the Search's criteria, if appropriate.

3. Click Find Now.

   All shortcuts are removed from the search's current list of results. As the Catalog finds items that satisfy the new search criteria, shortcuts to those items are added to the search results list.

4. Click the Close button to hide the Search dialog box.

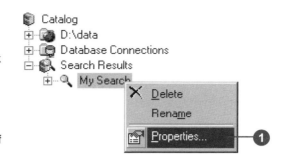

Deleting a shortcut

1. Click a shortcut in the Catalog tree.

2. Click the Delete button.

   The shortcut is removed from the search results list.

Selecting a shortcut's original item

1. Right-click a shortcut in the Catalog tree and click Go To Target.

   The original item is selected in the Catalog tree.

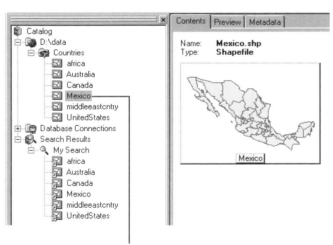

The shortcut's target is selected in the Catalog tree.

Working with maps and layers

11

With ArcMap, you can build, design, query, and analyze the data in your maps. The role of ArcCatalog is to help you find maps and locate data that you want to add to them. When you add data to a map, either by dragging and dropping from the Catalog or by using the Add Data dialog box, ArcMap creates a layer that references the source data. After refining a layer's labels and symbology, you can save it as a separate file so that you can use it again in other maps or e-mail it to someone else along with the data. You can also create layer files directly in ArcCatalog.

If you're managing an ArcSDE geodatabase, you can create layers for others to use and place them in a shared folder on the network. If different departments use different data in their work, place customized layers in separate folders. The rest of the organization can use those layers without having to know which tables in the database contain what data or how their attributes are related to your geographic features. With layers, people can focus on what the data means rather than how to access it.

Opening a map

When browsing through a folder's contents, you might find a map document or template that you want to work with. After drawing the map's contents and looking at its metadata to ensure it's the right map, you can open it by double-clicking it in the Catalog. In ArcMap, you can draw, query, edit, or analyze the map's contents.

If you want to create a new map, click the Launch ArcMap button on the Standard toolbar, choose to create a new map, and then start adding your data to it. If you want to use a template, simply click on the template that you would like to use. Templates can be helpful since they are ready to use.

Tip
Launching ArcToolbox
You can start ArcToolbox in the same way that you can start ArcMap from ArcCatalog. Click Launch ArcToolbox on the Standard toolbar.

See Also
For detailed information about working with maps, see Using ArcMap.

Opening an existing map

1. Navigate to the map you want to open.

2. Double-click the map either in the Contents list or in the Catalog tree.

   The map opens in ArcMap.

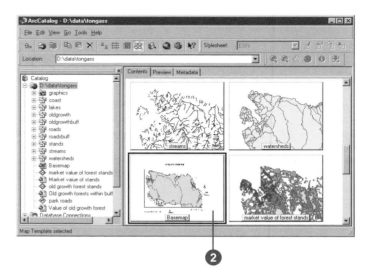

Creating a new map

1. Click Launch ArcMap on the Standard toolbar.

2. Click A new empty map and click OK.

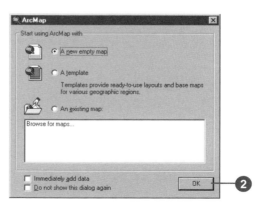

Adding data to maps

With the Catalog, you can locate the layers or data sources that you want to use and then drag and drop them onto the map. You can drop the data onto the data view or layout view in the ArcMap window. You can also drop geographic data sources directly into the correct spot in the map's table of contents. If you drop a table onto the map, it will only appear in the Source tab of the map's table of contents. If the ArcCatalog window or another application is hiding the ArcMap window, drag the data to the Windows taskbar and hold the mouse pointer over the ArcMap program icon for a second or two. After the ArcMap window appears on top, drop the data onto the map.

1. In ArcMap, open the map to which you want to add data.

2. In the Catalog, navigate to the data source you want to add to the map.

3. Drag the data from the Catalog.

4. Drop the data on the data view or layout view or in the map's table of contents.

   The new data appears in the map.

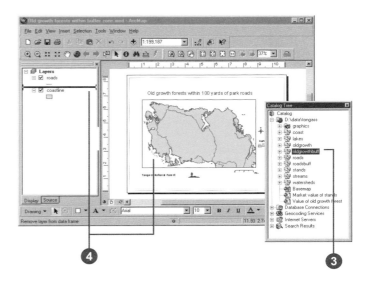

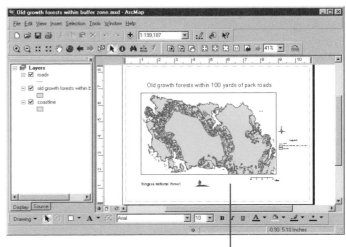

The new geographic data source appears in the map.

Creating layers

It takes time to analyze a feature class's attributes and symbolize its features so that people will readily understand the information that lies within the data. Layers that are created in ArcMap are stored within the map document. After finalizing the symbolization and labeling of a layer, you can save it outside the map as a separate layer file. You can reuse those layers in other maps or e-mail them along with the data to other people.

If you're managing a multiuser geodatabase, you can create layers for other people to use without having to open ArcMap. You might create a series of layers from the same feature class representing different feature attributes such as age or income and then share them on the network. Others can drop those layers onto their maps without having to know how to access the database.

Sometimes you'll want to group several geographic data sources together in a layer because you want to manage them as one entry in the map's table of contents. For example, you might group all of a map's background layers together, or you may want to group several layers storing transportation features such as ▶

Saving a layer outside a map

1. In ArcMap, open the map containing the layer that you want to save as a layer file.

2. Right-click the layer in the map's table of contents.

3. Click Save As Layer File.

4. Navigate to the folder in which you want to save the layer.

5. Type a name for the layer file.

6. Click Save.

7. In the Catalog, navigate to the folder in which you saved the layer.

   The layer is now an item in the folder. If the layer doesn't appear in the Contents list, click the folder in which it should appear in the Catalog tree, click the View menu, then click Refresh to update the list.

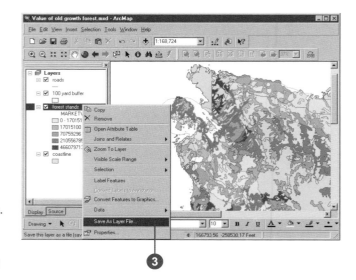

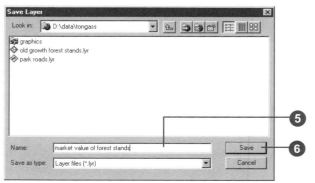

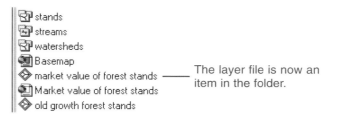

The layer file is now an item in the folder.

roads, highways, railways, and ferries. You can change which layers belong to the group from the group layer's Properties dialog box. A group layer can combine data from several data sources, each of which may store data in a different format; for example, you can combine TIN, coverage, and shapefile data in the same group layer.

Creating a new layer

1. Select the folder in which you want to store the new layer in the Catalog tree.

2. Click the File menu, point to New, then click Layer.

3. Type a name for the new layer.

4. Click the Browse button.

5. Navigate to and click the geographic data source for which you want to create the layer.

6. Click Add.

7. If you don't want the Catalog to create a thumbnail representing the entire layer, uncheck Create thumbnail.

8. If you don't want the layer to store the full pathname identifying the location of the data, check Store relative path name. The location of the data will be recorded in relation to where the layer itself is stored.

9. Click OK.

   The new layer appears in the folder's contents.

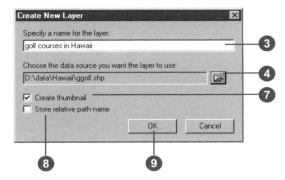

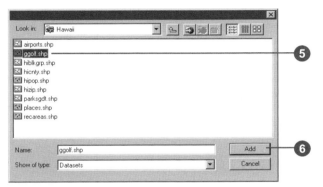

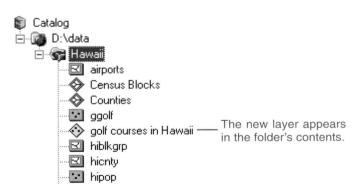

The new layer appears in the folder's contents.

Creating a layer from the data

1. Right-click the data source from which you want to create a layer.

2. Click Create Layer.

3. Navigate to the folder in which you want to save the layer.

4. Type a name for the layer file.

5. Click Save.

   The layer file appears in the folder's contents.

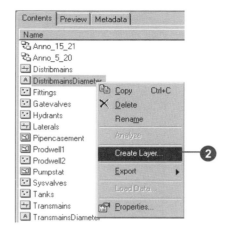

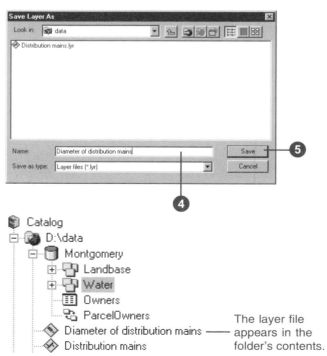

The layer file appears in the folder's contents.

Creating a new group layer

1. Select the folder in which you want to place the new group layer in the Catalog tree.

2. Click the File menu, point to New, then click Group Layer.

3. Type a name for the new group layer.

4. Press Enter.

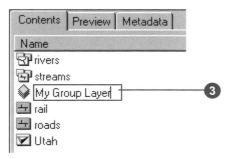

Adding layers to a new group layer

1. Right-click a group layer, click Properties, then click the Group tab.

2. Click Add.

3. Navigate to and click the geographic data source or layer that you want to add to the group. Click Open.

4. Repeat step 3 until all layers have been added.

5. Layers are drawn in order from the bottom to the top of the list. Click a layer and then click the arrow buttons to reorder the list.

6. Click a layer and click Properties to set the properties of individual layers.

7. Click OK.

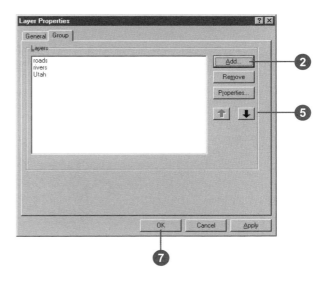

Creating a group layer from the data

1. Hold down the Shift or Ctrl key and click the appropriate geographic data sources in the Contents tab. The data sources must all have the same format—for example, select three raster datasets.

2. Right-click one of the selected data sources.

3. Click Create Layer.

4. Navigate to the folder in which you want to save the group layer.

5. Type a name for the group layer file.

6. Click Save.

   The new group layer appears in the folder's contents.

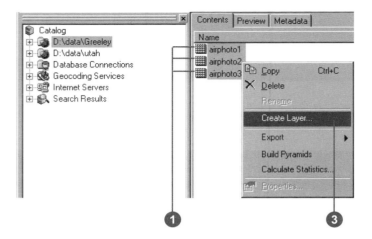

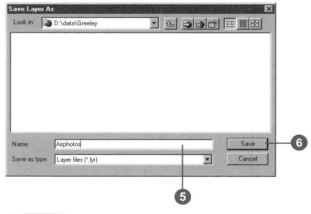

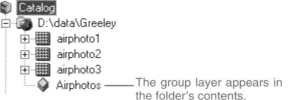

The group layer appears in the folder's contents.

Setting a layer's properties

The Layer Properties dialog box will be different for different types of geographic data. For example, defining the symbology of a vector data source will be different than for a raster dataset. If you have write permission for the data, you can change the layer's properties such as its symbology. With a group layer, some properties apply to the whole group, but you can also edit the properties of each of its layers individually.

One property common to all layers is their data source. Layer files identify their data sources using a path. If the folder connection through which the data was accessed connects directly to a local disk drive, the path in the layer file will use the local disk name such as C:\. If you send this layer to others, they can't access your data and will be unable to preview its contents. If the folder connection was created from the Network Neighborhood, the path will include the names of the computer and the Share Name such as \\Blues\SharedData. Others will be able to access data stored on your computer if the layer's path accesses the ▶

Setting a layer's properties

1. Right-click the layer whose properties you want to set.

2. Click Properties.

3. Set the layer's properties in the Layer Properties dialog box.

4. Click OK.

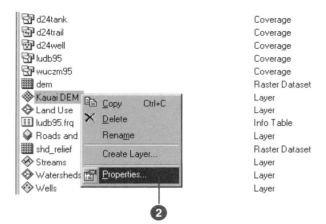

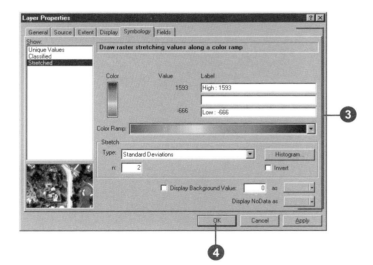

data using this method. However, if you rename or move the data, you must update the layer files that access that data.

Another alternative for referencing a layer's data source is to use a relative pathname. Suppose a folder named forest contains both a layer and another folder named data. The layer's data source is located within the data folder. With a relative path, the layer will start looking for the data source from the location in which the layer is stored. With the relative path "data/myShapefile", the layer will be able to find its data, even if the forest folder is relocated or renamed. The layer will only need to be repaired if the data folder or the data source itself is moved or renamed. You only have the option to use relative paths when you create a new layer in the Catalog.

You can label, query, or symbolize a layer's data using attributes stored in another table. To do so, define the relationship between the layer's data and the table in the Layer Properties dialog box. If you join the table, its attributes are appended to the layer's data. If you establish a relate instead of a join, you can explore the related attributes in ArcMap, but you can't use them to set the ▶

Repairing a layer

1. Right-click the layer you want to fix and click Properties.

2. Click the Source tab.

   When a layer can't find its data source, you'll see information about the data it's looking for but no extent or coordinate system information.

3. Click Set Data Source.

4. Navigate to and click the layer's data source in the Browse dialog box. Click Add.

   The Source tab now shows the path to the data source as well as its extent and the properties of its coordinate system.

5. Click OK.

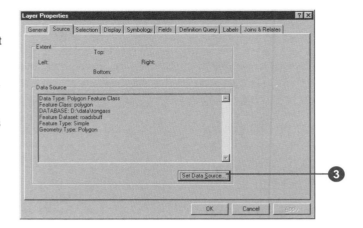

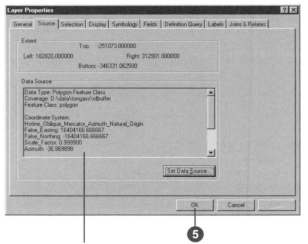

When the layer can access its data source, you will see extent and coordinate system information in the Source tab.

layer's properties. Any two data sources can be joined or related through a layer regardless of their format as long as they share a common attribute. For example, you might join attributes in a geodatabase table to a shapefile's features based on a common attribute such as "CustomerID". Relationship classes define the relationships among coverages and INFO tables in a folder or among feature classes and tables in a geodatabase. When a relationship class exists, the layer can use its information to join the two related data sources.

See Also

For detailed information about setting layer's properties, see Using ArcMap.

Joining attributes to the layer's data

1. Right-click the layer to which you want to join attributes and click Properties.

2. Click the Joins & Relates tab.

3. Click Add next to the Joins list.

4. Click the dropdown arrow and click Join attributes from a table.

5. Click the dropdown arrow and click the attribute that will be used to join the layer's data to an external data source.

6. Click the Browse button. Navigate to and click the data source whose attributes will be joined to the layer, then click Add.

7. Click the dropdown arrow and click the attribute in the external data source that contains the same values as the attribute chosen in step 5.

8. Click OK.

   The data source's name is added to the list of those that have been joined to the layer. You can now label, query, and symbolize the layer's data using attributes in the joined data source.

9. Click OK.

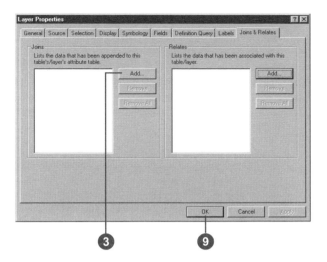

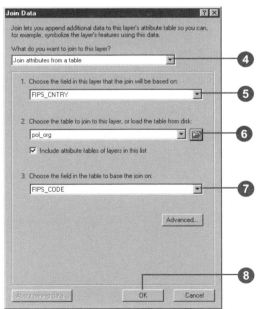

Joining attributes using a relationship class

1. Right-click the layer to which you want to join attributes. Click Properties.

2. Click the Joins & Relates tab.

3. Click Add next to the Joins list.

4. Click the dropdown arrow and click Join data based on a pre-defined relationship class.

5. Click the dropdown arrow and click the name of the relationship on which the join will be based.

6. Click OK.

   The data source's name is added to the list of those that have been joined to the layer. You can now label, query, and symbolize the layer's data using attributes in the joined data source.

7. Click OK.

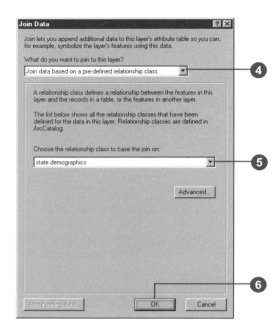

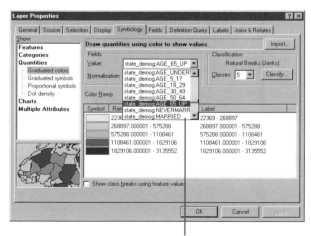

You can symbolize the layer's data using joined attributes.

Working with shapefiles

12

The Catalog lets you create new shapefiles and dBASE tables and modify them by adding, deleting, and indexing attributes. You can also define a shapefile's coordinate system and update its spatial index. While you can change the structure and properties of a shapefile in ArcCatalog, you must use ArcMap to modify its features and attributes—for example, to add values into a new column by performing a calculation on values in another column.

Creating new shapefiles and dBASE tables

New shapefiles and dBASE tables can be created in ArcCatalog. When you create a new shapefile, you must define the type of features it will contain, whether those features will represent routes, and whether those features will be three-dimensional. These properties can't be modified after the shapefile has been created. If you choose to define the shapefile's coordinate system later, until then it will be classified as "Unknown".

The process of defining the new shapefile or dBASE table's attributes is separate from creating the shapefile or dBASE table itself. After creating the item, right-click it in the Catalog and click Properties to define its attributes. Because they must contain at least one column, the Catalog adds a default column to the shapefile or dBASE table when it is created. For shapefiles, an integer column named "Id" is added as an attribute. For dBASE tables, the text column "Name1" is added. Add the appropriate attributes to your shapefile or dBASE table, then delete the default column.

Creating a new shapefile

1. Select a folder or folder connection in the Catalog tree.

2. Click the File menu, point to New, and click Shapefile.

3. Click in the Name text box and type a name for the new shapefile.

4. Click the Feature Type dropdown arrow and click the type of feature the shapefile will contain.

5. Click Edit to define the shapefile's coordinate system.

6. In the Spatial Reference Properties dialog box, click Select and choose a predefined coordinate system.

   Or click Import and choose the data source whose coordinate system you want to copy.

   Or click New and define a new, custom coordinate system.

7. Click OK. ▶

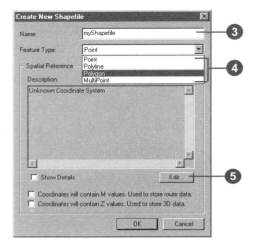

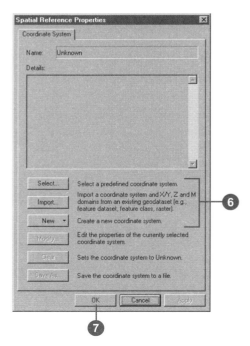

8. If the shapefile will store polylines representing routes, check Coordinates will contain M values.

9. If the shapefile will store three-dimensional features, check Coordinates will contain Z values.

10. Click OK.

The new shapefile appears in the folder's Contents.

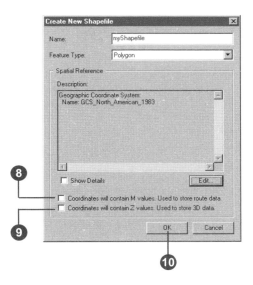

Creating a new dBASE table

1. Select a folder or folder connection in the Catalog tree.

2. Click the File menu, point to New, and click dBASE Table.

A new dBASE table appears in the folder's Contents.

3. Type a new name for the table and press Enter.

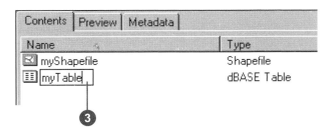

Adding and deleting attributes

The Catalog lets you modify the structure of shapefiles and dBASE tables by adding and deleting attribute columns. The name and data type of an existing column can't be modified; instead, you must add a new column with the appropriate name and data type. A column's name must be no more than 10 characters in length; additional characters will be truncated. A shapefile's FID and Shape columns and a dBASE table's OID column can't be deleted. The OID column is a virtual column created by ArcGIS software when accessing the table's contents; it guarantees that each record in the table has at least one unique value. Shapefiles and dBASE tables must have at least one attribute column in addition to the FID and Shape columns or the OID column. After adding attributes, you must start an edit session in ArcMap to define their values.

Adding an attribute

1. Click the shapefile or dBASE table to which you want to add an attribute.

2. Click the File menu and click Properties.

3. Click the Fields tab.

4. Scroll down until you see the last attribute.

5. Click in the empty row below the last attribute under Field Name and type the name of the new attribute.

6. Click under Data Type next to the new attribute name. A dropdown list appears with the Text data type selected by default. Click the appropriate data type in the list for the new attribute.

   The properties that are appropriate to the new attribute's data type appear in the Field Properties list below.

7. Click in the Field Properties list and type the properties for the new attribute.

   For example, if the data type is a real number, set the precision (the total number of digits the values can have) and scale (the total number of decimal places the values can have) properties.

8. Click OK.

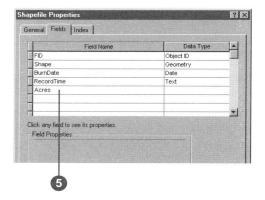

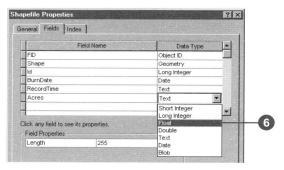

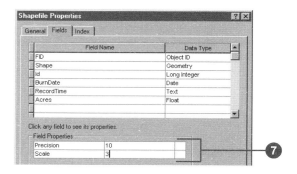

Deleting an attribute

1. Click the shapefile or dBASE table whose columns you want to delete.

2. Click the File menu and click Properties.

3. Click the Fields tab.

4. Position the mouse pointer over the gray button to the left of the column you want to delete. The pointer's icon changes to an arrow.

5. Click the gray button to select the column.

   The column is selected and its properties appear in the Field Properties list below.

6. Press the Delete key on the keyboard.

   The selected attribute is removed from the list of columns.

7. Click OK.

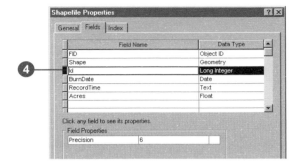

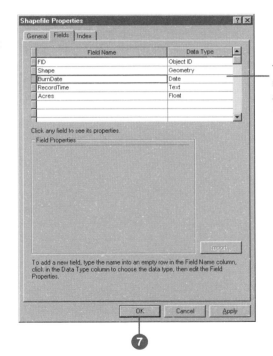

The attribute no longer appears in the list of columns.

Creating and updating indexes

With ArcCatalog you can add attribute indexes to shapefiles and dBASE tables. An index can improve the performance of queries that evaluate an attribute's values. Indexes created by ArcCatalog and used by ArcGIS software are different from those created and used by ArcView GIS 3. Attribute indexes created in ArcCatalog are automatically maintained as values in the column change.

In addition to letting you add attributes, the Catalog also lets you add, update, and delete a shapefile's spatial index. As features are added to or removed from the shapefile, its spatial index will be updated automatically. However, at times you may want to update the shapefile's spatial index manually; this process also updates the extent of its features. Having a current spatial index ensures that a high level of performance is maintained when drawing and working with the shapefile's features.

Indexing an attribute

1. Click the shapefile or dBASE table to which you want to add an attribute index.

2. Click the File menu and click Properties.

3. Click the Index tab.

4. Check an attribute to index its values.

   Uncheck an attribute to delete its index.

5. Click OK.

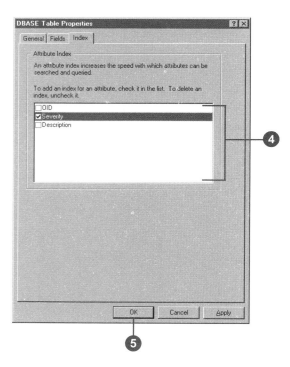

Adding a spatial index

1. Click the shapefile for which you want to create a spatial index.

2. Click the File menu and click Properties.

3. Click the Index tab.

4. Click Add.

5. Click OK.

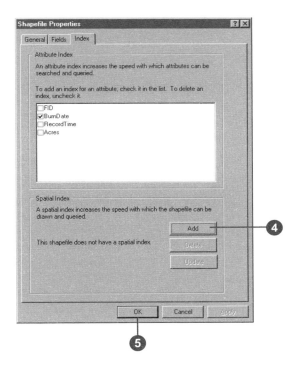

Updating a spatial index

1. Click the shapefile to which you want to add a spatial index.

2. Click the File menu and click Properties.

3. Click the Index tab.

4. Click Update.

5. Click OK.

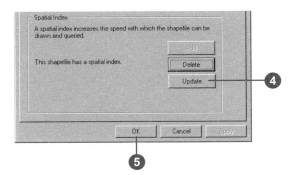

Defining a shapefile's coordinate system

A shapefile often doesn't have any information defining which coordinate system was used to define its features. In this case, the Shape column's Spatial Reference property will be "unknown" or "assumed geographic". If the features' bounding coordinates are within the range of -180 and 180 in the x direction and -90 and 90 in the y direction, ArcGIS software assumes the data to be geographic and its datum to be NAD27. You can work with shapefiles even if their coordinate system hasn't been defined, but you may not be able to take advantage of all the available functionality. For example, you may be unable to add the shapefile to some maps, and its automatically created metadata will be incomplete.

You can define a shapefile's coordinate system in ArcCatalog in several ways. You can select one of the predefined coordinate systems provided with ArcCatalog, import the coordinate system parameters used by another data source, or define a new, ▶

Defining a shapefile's coordinate system

1. Click the shapefile whose coordinate system you want to define.

2. Click the File menu and click Properties.

3. Click the Fields tab.

4. Click the Shape column in the column list.

5. In the Field Properties list below, click the ellipses button (...) next to the Spatial Reference property.

6. In the Spatial Reference Properties dialog box, click Select and then choose a predefined coordinate system.

   Or click Import and then choose the data source whose coordinate system parameters you want to copy.

   Or click New, click Geographic or Projected, and then define a new, custom coordinate system.

7. Click OK in the Spatial Reference Properties dialog box. ▶

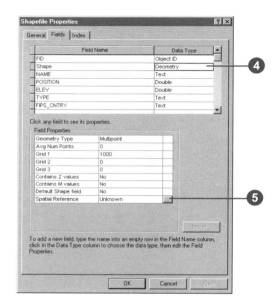

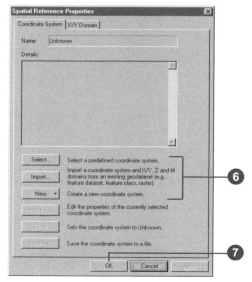

custom coordinate system. Afterwards, you can modify individual parameters as required. A shapefile's coordinate system parameters are stored in the same location as the shapefile, in a .prj file with the same name as the shapefile.

Once a coordinate system has been defined, you can modify individual parameters. For example, you might want to modify one parameter in a coordinate system that was imported from another data source or customize one of the predefined coordinate systems. After creating a custom coordinate system, you can save it as a separate file. You might want to share your coordinate system with others in your organization.

In the Shapefile Properties dialog box, the name of the coordinate system appears next to the Spatial Reference property in the Field Properties list.

8. Click OK in the Shapefile Properties dialog box.

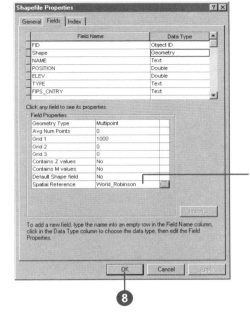

The coordinate system's name appears next to the Spatial Reference property.

Tip

Defining coordinate systems of other data sources

The way you define a shapefile's coordinate system is the same way you define coordinate systems for feature classes in geodatabases, CAD drawings, and rasters that aren't grids. For grids and TIN datasets, follow the steps for defining a coverage's coordinate system in Chapter 13, 'Working with coverages'.

Selecting an existing coordinate system

1. In the Spatial Reference Properties dialog box, click Select. ▶

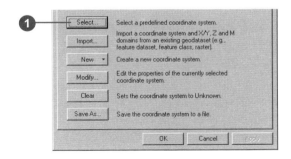

2. Navigate to the coordinate system you want to use. For example, you can use one of the predefined coordinate systems in the Coordinate Systems folder that was provided with ArcCatalog. Or you can use a projection file that was created with ArcInfo Workstation.

3. Click the coordinate system.

4. Click Add.

   The coordinate system's parameters are listed in the Spatial Reference Properties dialog box.

5. If you wish, click Modify to change the coordinate system parameters. Or click Clear and repeat steps 1 through 4.

6. Click OK in the Spatial Reference Properties dialog box.

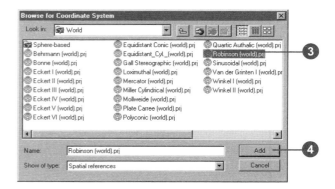

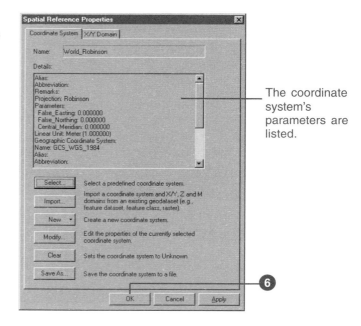

The coordinate system's parameters are listed.

Importing a coordinate system

1. In the Spatial Reference Properties dialog box, click Import.

2. Navigate to the data source whose coordinate system parameters you want to copy. For example, you can get coordinate system information from coverages, rasters, or feature datasets and feature classes in a geodatabase.

3. Click the data source.

4. Click Add.

   The coordinate system's parameters are listed in the Spatial Reference Properties dialog box.

5. If you wish, click Modify to change the coordinate system parameters. Or click Clear and repeat steps 1 through 4.

6. Click OK in the Spatial Reference Properties dialog box.

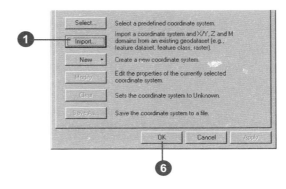

Modifying default parameters

When creating a custom coordinate system, if you choose an object such as a datum or spheroid from one of the dropdown lists, those parameter values will be read-only. To modify them, click the dropdown arrow again and click "<custom>". Change the default values appropriately and then type a name for your custom set of parameters.

Defining a new geographic coordinate system

1. In the Spatial Reference Properties dialog box, click New and click Geographic.

2. Type a name for the custom coordinate system.

3. Type the appropriate semi-major and semiminor or inverse flattening values and type a name for your custom Spheroid and Datum.

   Or click the Datum or Spheroid Name dropdown arrow and click a predefined datum or spheroid.

4. Type the appropriate Radians per unit and type a name for your custom units.

   Or click the Angular Unit dropdown arrow and click a predefined unit of measure.

5. Type the appropriate degrees, minutes, and seconds defining the prime meridian and type a name for this line of longitude.

   Or click the Prime Meridian dropdown arrow and click a predefined line of longitude.

6. Click OK.

7. Click OK in the Spatial Reference Properties dialog box.

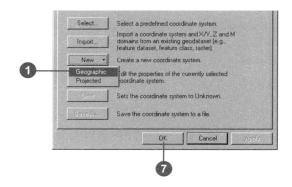

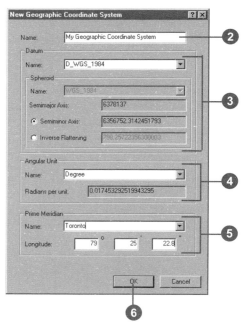

Defining a new projected coordinate system

1. In the Spatial Reference Properties dialog box, click New and click Projected.

2. Type a name for the custom coordinate system.

3. Click the Projection Name dropdown arrow and click one of the supported projections.

4. Type the appropriate parameter values for the projection.

5. Click the Linear Unit dropdown arrow and click a predefined unit of measure.

   Or click "<custom>" in the dropdown list, type the appropriate meters per unit, then type a name for your custom units.

6. Define the projection's datum by selecting a predefined geographic coordinate system or defining a new geographic coordinate system. Afterwards, you can modify the geographic coordinate system's parameters if you wish.

7. Click OK.

8. Click OK in the Spatial Reference Properties dialog box.

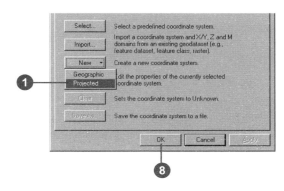

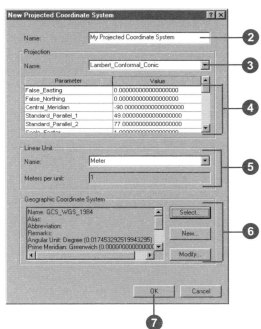

Modifying a coordinate system's parameters

1. In the Spatial Reference Properties dialog box, click Modify.

   The Geographic or Projected Coordinate System Properties dialog box appears.

2. Follow the steps for defining either a new geographic or a new projected coordinate system to change the appropriate parameters.

3. Click OK.

4. Click OK.

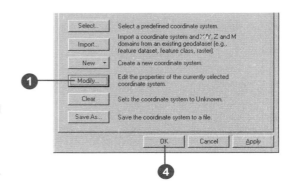

Saving a coordinate system to a file

1. In the Spatial Reference Properties dialog box, click Save As.

2. Navigate to the location where you want to place the coordinate system file. For example, place it in a shared folder on the network.

3. Type a name for the coordinate system file and click Save.

4. Click OK.

   The coordinate system file appears in the folder's Contents list.

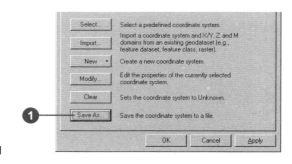

The coordinate system appears in the folder's Contents list.

Working with coverages

13

IN THIS CHAPTER

- **Creating a new coverage**

- **Creating a new INFO table**

- **Generating topology**

- **Defining a coverage's coordinate system**

- **Modifying a coverage's tics and extent**

- **Setting a coverage's tolerances**

- **Maintaining attributes**

- **What is a relationship class?**

- **Creating a coverage relationship class**

When ArcInfo is installed on your computer, you can use ArcCatalog to manage your coverages. You'll know ArcInfo is installed on your computer if "ArcInfo" appears in the title bar of the ArcCatalog window. ArcCatalog lets you build a coverage's topology, define its projection, and set its tolerances. Similarly, ArcCatalog lets you add, alter, and index attributes for coverage feature classes and INFO tables.

With ArcCatalog, you can also define the associations between coverages and INFO tables in a folder by creating relationship classes. When editing a coverage's features in ArcMap, if the coverage participates in a relationship class with an INFO table, you can edit the INFO table's attributes at the same time. In ArcCatalog, if there is a composite relationship between coverages, then when a feature is moved, its related feature is moved in the same way.

ArcInfo and ArcEditor

Creating a new coverage

ArcCatalog lets you create new, empty coverages. When assembling data, it's important to use a master coverage as a template for all the coverages in a project so that they can be overlaid correctly; otherwise, common features in different coverages, such as a shoreline, may not line up. When you create a new coverage using a template, the template coverage's tic points, boundary, and coordinate system information are copied to the new coverage. If you don't use a template, you must add tics to the new coverage before you can add features to it. You aren't required to set the new coverage's boundary before adding features to it.

The New Coverage Wizard prompts you to choose whether the new coverage will contain single- or double-precision coordinates. You must also choose which feature class will have topology; the wizard will create the appropriate feature classes and feature attribute tables in the new coverage based on your choice.

1. In the Catalog tree, click the folder in which the new coverage will be created.

2. Click the File menu, point to New, and click Coverage.

3. Type a name for the new coverage.

4. Check the box to use another coverage as a template.

5. Click the Browse button and then navigate to the coverage that will be used as a template. Click the coverage and click Open.

6. Click Next.

7. If you want to define or modify the coordinate system information for the new coverage, click Define, then follow steps 5 through 12 in the task 'Defining a coordinate system interactively'.

8. Click Next.

9. Click the dropdown arrow and click the feature class that will have topology.

10. Click Single if you want to create a single-precision coverage. Double-precision is the default.

11. Click Finish.

   The new coverage appears in the Catalog.

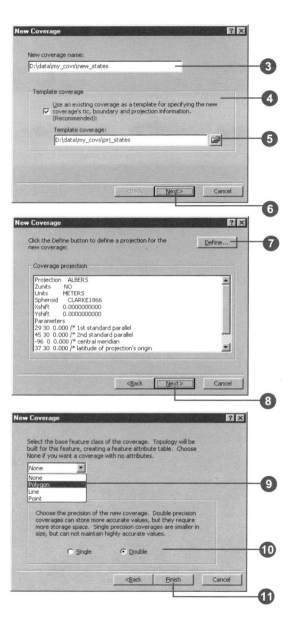

Creating a new INFO table

You can create new, empty INFO tables in ArcCatalog. The table's name must be 32 characters or less, while column names must not exceed 16 characters in length. Columns are defined using standard ArcInfo data types such as Binary. The input width is the maximum number of characters or bytes used to store the column's values. For numeric columns, the width must be large enough to accommodate the decimal point and negative sign. The display width is the number of spaces used to display values in ArcInfo Workstation; for decimal values, the display width should be one space greater than the input width to account for the decimal point.

1. In the Catalog tree, click the folder in which the new INFO table will be created.
2. Click the File menu, point to New, and click INFO table.
3. Type a name for the new table.
4. Click the data type of the first column in the table.
5. Type a name for the column.
6. Default values for the column's width, display width, and number of decimal places are provided based on the column's data type. You can change these values if you wish.
7. Click New item to add another column to the table, then repeat steps 4 through 6 to define the new column.
8. Repeat step 7 until all columns have been added to the table.
9. Use the arrows to navigate back and forth through the table's columns to review and modify their properties, if necessary. Click Remove item if a column must be deleted.
10. Click OK.

    The new INFO table appears in the Catalog.

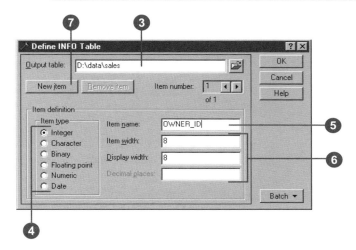

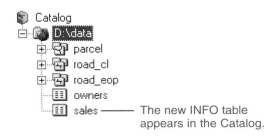

The new INFO table appears in the Catalog.

ArcInfo and ArcEditor

Generating topology

The General tab in the Coverage Properties dialog box provides important information about a coverage. In addition to showing which feature classes have topology, you can find out where the coverage is stored on disk and whether it is a single- or double-precision coverage. When you click a feature class, the number of features it contains appears at the bottom of the tab.

If topology is missing for a feature class that should have it, you can generate topology using either the Build or Clean commands in the Coverage Properties dialog box. You might also use the Build command to create a feature attribute table for a feature class. Build assumes the coordinate data is correct, while Clean finds arcs that cross and places a node at each intersection. Clean also corrects under-shoots and overshoots within a specified tolerance. For polygon and region coverages with preliminary topology, a red warning indicator appears in the icons for both the coverage and the appropriate feature class.

Building coverage topology

1. Right-click the coverage for which you want to build topology and click Properties.

2. Click the General tab.

3. Click the feature class for which you want to build topology.

4. Click Build.

5. Change the feature class or annotation class for which to build topology, if appropriate.

6. Click OK in the Build dialog box.

7. Click OK.

A coverage with preliminary topology can be identified by its icon.

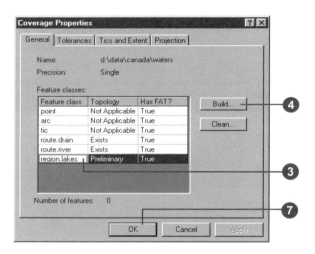

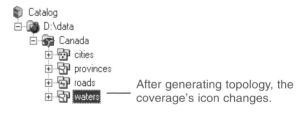

After generating topology, the coverage's icon changes.

Tip

Setting tolerances

You can set the fuzzy and dangle tolerances and any other tolerance for the coverage in advance using the Tolerances tab of the Coverage Properties dialog box. See 'Setting a coverage's tolerances' later in this chapter.

Cleaning a coverage

1. Right-click the coverage that you want to clean and click Properties.

2. Click the General tab.

3. Click the feature class that you want to clean.

4. Click Clean.

5. Type in the fuzzy and dangle tolerances in the Clean dialog box, if appropriate.

6. If necessary, check Clean lines only.

7. Click OK in the Clean dialog box.

8. Click OK.

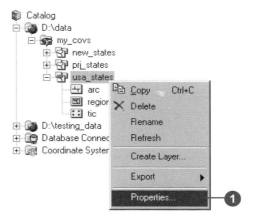

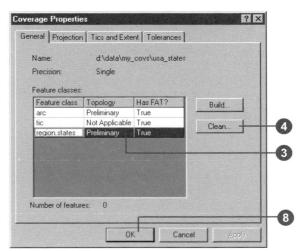

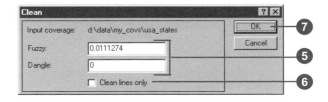

Defining a coverage's coordinate system

The Projection tab in the Coverage Properties dialog box shows the coverage's coordinate system and lists its parameters. The coordinate system defines how coordinates describing features on the earth's surface are mathematically transformed to accurately represent them on a flat map sheet. If the coverage's coordinate system hasn't been defined, you can do so from the Coverage Properties dialog box. The Define Projection Wizard guides you through the process.

ArcCatalog lets you either define the coordinate system's parameters interactively or copy the information from another coverage, a grid, or a TIN. If the coverage already has coordinate system information, you can also use the Define Projection Wizard to change that information.

Matching another item's coordinate system

1. Click the coverage whose coordinate system you want to define.

2. Click the File menu and click Properties.

3. Click the Projection tab.

4. Click Define.

5. Click Define a coordinate system for my data to match existing data.

6. Click Next. ►

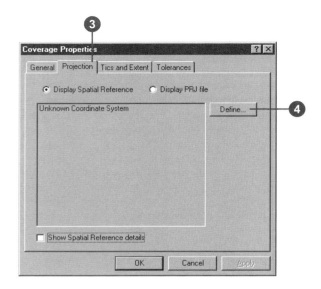

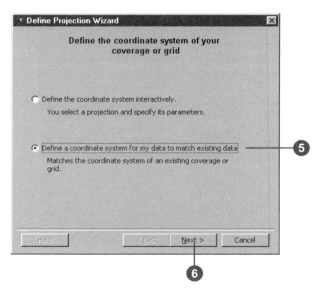

Tip

Different ways to see projection information

In the Projection tab, there are two different ways to look at a coverage's projection information. When you click Display PRJ file, you see exactly how the projection is defined in the coverage's PRJ file. When you click Display Spatial Reference, you see the Projection Engine's interpretation of the information in the PRJ file; check Show Spatial Reference details to list all of the coordinate system's parameters. The Spatial Reference version is the one that is recorded in a coverage's metadata; this way coordinate system information will be recorded in the same format for all geographic data sources.

Tip

Defining coordinate systems for grids and TINs

The way you define the coordinate system of a raster dataset that is a grid and a TIN dataset is the same way you define a coordinate system for a coverage. For other rasters, follow the steps for defining a shapefile's coordinate system.

7. Click the Browse button.

8. Navigate to and click the coverage, grid, or TIN whose coordinate system you want to use. Click Open.

9. Review the coordinate system's parameters that appear in the wizard. If you want to use this coordinate system, click Next. Otherwise, locate a different coverage.

10. Click Finish.

    The coordinate system and its parameters now appear in the Coverage Properties dialog box.

11. Click OK.

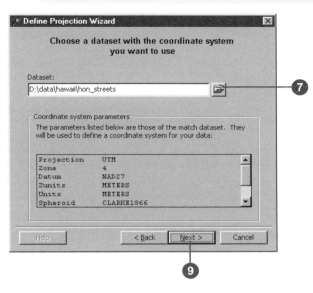

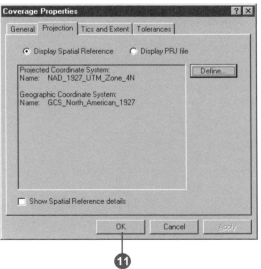

ArcInfo and ArcEditor

Defining a coordinate system interactively

1. Click the coverage whose coordinate system you want to define.

2. Click the File menu and click Properties.

3. Click the Projection tab.

4. Click Define.

5. Click Next.

6. Click the appropriate coordinate system in the Projections list on the left.

   A description of the coordinate system appears on the right.

7. Click Next. ▶

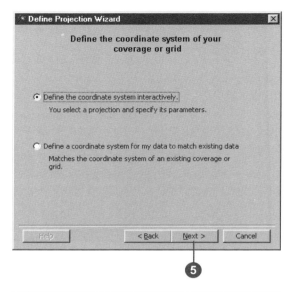

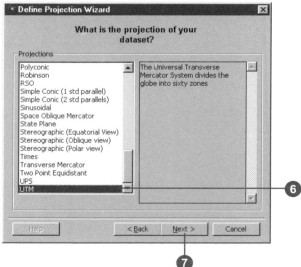

8. Enter the appropriate parameter values for the coordinate system; each one has a different set of parameters.

9. Click Next.

10. If the coverage uses a projected coordinate system, click the datum it uses or click Spheroid and define the spheroid's parameters.

11. Click Next.

12. Review the summary of the coordinate system that will be assigned to the coverage. If you want to modify the coordinate system's parameters, go back through the wizard by clicking the Back button. Click Finish if you want to use this coordinate system.

The coordinate system and its parameters now appear in the Coverage Properties dialog box.

13. Click OK.

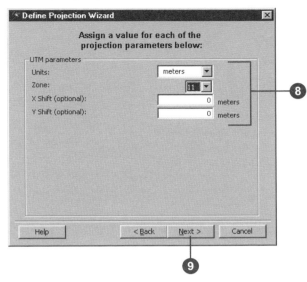

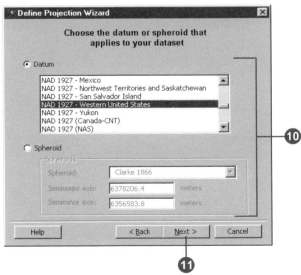

ArcInfo and ArcEditor

Modifying a coverage's tics and extent

Every coverage has a set of tic points as well as boundary, or extent, information. Tics are points on a map sheet for which real-world coordinates are known. Before a digitizing session, you can register tics from the paper map with the digitizer and existing features in the coverage. Later, you can use the tics to transform feature coordinates from digitizer units to a known coordinate system. From the Coverage Properties dialog box, you can add or update tics if you know their coordinates. You can also delete tics from the coverage.

A coverage's extent defines the maximum and minimum x,y coordinate values in the coverage. Usually, ArcInfo maintains the extent for you, updating it as you add features to or remove features from a coverage. If for some reason the extent shown does not match the actual extent of the features, click Fit in the Coverage Properties dialog box to recalculate the coverage's extent. Alternatively, you can type new extent values into the appropriate text boxes.

Adding tics

1. Right-click the coverage to which you want to add a tic and click Properties.

2. Click the Tics and Extent tab.

3. Click Add.

4. Double-click in the X column and type the tic point's x-coordinate.

5. Double-click in the Y column and type the tic point's y-coordinate.

6. Click OK.

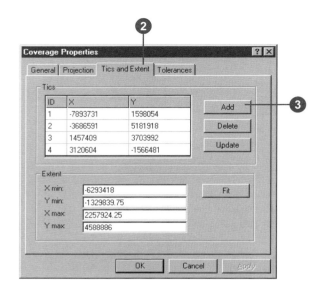

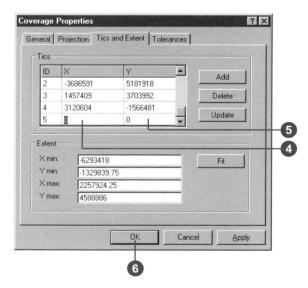

Updating tic coordinates

1. Right-click the coverage whose tics you want to modify and click Properties.

2. Click the Tics and Extent tab.

3. Click the ID of the tic whose coordinates you want to modify.

4. Click Update.

5. Click in the X column and type the tic point's x-coordinate.

6. Double-click in the Y column and type the tic point's y-coordinate.

7. Click OK.

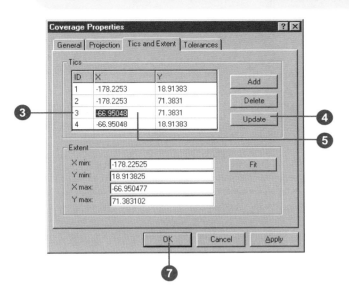

Deleting tics

1. Right-click the coverage whose tics you want to delete and click Properties.

2. Click the Tics and Extent tab.

3. Click the ID of the tic you want to delete.

4. Click Delete.

5. Click OK.

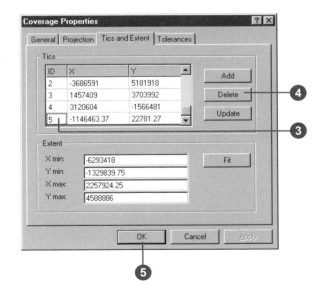

ArcInfo and ArcEditor

Recalculating the extent

1. Right-click the coverage whose extent you want to recalculate and click Properties.

2. Click the Tics and Extent tab.

3. Click Fit.

4. Click OK.

   The extent is recalculated.

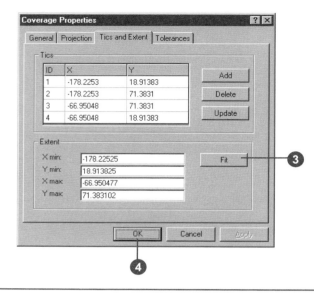

Updating the extent manually

1. Right-click the coverage whose extent you want to modify and click Properties.

2. Click the Tics and Extent tab.

3. Click in the text box of the appropriate extent value and type the new coordinate value.

4. Repeat step 3 until all the extent coordinates are correct.

5. Click OK.

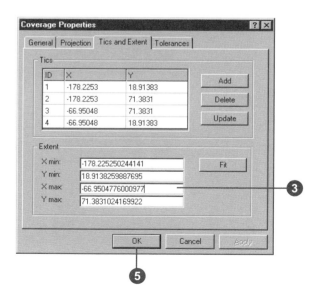

Setting a coverage's tolerances

Many coverage operations use tolerances. For example, when you digitize or edit with node snapping turned on, two nodes are automatically snapped together if they fall within a given distance of each other; that distance is the Node Snap tolerance. Tolerances are distances in the coverage's units. If the feature coordinates are in digitizer units, the tolerances are likely in inches; if the features are projected, the tolerances are likely in feet or meters.

The Tolerances tab lists the current values for all of a coverage's tolerances. You'll see "Default" next to values that have been suggested by ArcInfo. You may want to change the default tolerances to values that are more appropriate for the type and scale of features in the coverage; when you do so in the Tolerances tab, "Default" disappears. You can also set tolerances when using an ArcToolbox tool or an ArcInfo Workstation command. When you open the Coverage Properties dialog box after running the tool or command, "Verified" appears next to the tolerance that was used.

1. Click the coverage whose tolerances you want to set.

2. Click the File menu and click Properties.

3. Click the Tolerances tab.

4. Click in the text box of the tolerance whose value you want to modify and type the new value.

5. Repeat step 4 until all the tolerance values are correct.

6. Click OK.

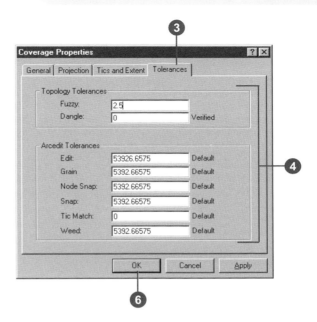

ArcInfo and ArcEditor

Maintaining attributes

Features usually have descriptive information stored in attributes. Attributes can either reside in the feature attribute table or in separate INFO tables; INFO tables can be associated with the feature attribute table using a relate or a relationship class. Both relates and relationship classes use a common attribute to establish a temporary connection between corresponding records in two tables.

The properties dialog box for an INFO table and a coverage feature class is the same. When you open the properties dialog box for a table or feature class, you will see an FID column, which contains the unique identifier for each record or feature. A feature class also has a Shape column and may have several *pseudo items* such as the angle of rotation for marker symbols. Pseudo items are maintained by ArcInfo; their names are preceded by a dollar sign ($), for example, $ANGLE. If a feature class has an attribute table, you will also see several attributes.

In the properties dialog box for a table or feature class, you can add, update, and delete ▶

Adding a new attribute

1. Right-click the coverage feature class or INFO table to which you want to add an attribute and click Properties.

2. Click the Items tab.

3. Under Item Name, click the attribute after which the new attribute should be placed. The attribute should not be a redefined attribute or a pseudo item. If an attribute is redefined, "Yes" appears in the Redefined column. For pseudo items, "N/A" appears under Column to the left of the attribute's name.

4. Click Add.

5. Type the new attribute's name in the Add Item dialog box.

6. Click next to Type, click the dropdown arrow, then click the appropriate data type for the values the attribute will contain. ▶

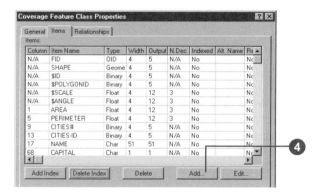

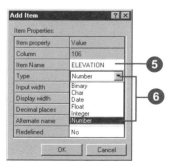

attributes and create indexes for the values the attributes contain. You can only modify and delete columns in the attribute table that are not maintained by ArcInfo. Index an attribute to improve the performance of operations that access its contents such as selecting features with specific attribute values. If you index the shape column, a spatial index will be created, which improves the performance of any operation that retrieves features by location. After modifying the values in a column, delete the existing index and then add a new one.

In addition to new attributes, you can add redefined attributes. Add a redefined attribute to combine or divide the values in existing attributes. Redefined attributes are useful for customizing how the values in a table are displayed. For example, you might choose to show one redefined attribute that concatenates values rather than showing two separate, adjacent columns.

7. Type an appropriate input width and display width for the attribute's values, then type the maximum number of decimals the values can have, if appropriate. Valid input widths may be presented as a dropdown list.

8. Type a more descriptive name for the attribute, if appropriate.

9. Click OK.

The new attribute appears in the Properties dialog box.

10. Click Apply to save your changes.

11. Click OK.

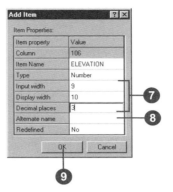

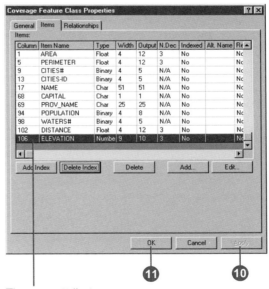

The new attribute appears in the Properties dialog box.

ArcInfo and ArcEditor

Adding a redefined attribute

1. Right-click the coverage feature class or INFO table to which you want to add a redefined attribute and click Properties.

2. Click the Items tab.

3. Under Item Name, click an attribute that is not a redefined attribute or a pseudo item. For redefined attributes, "Yes" appears in the Redefined column. For pseudo items, "N/A" appears under Column to the left of the attribute's name.

4. Click Add.

5. Click next to Redefined, click the dropdown arrow, and click Yes.

6. Click next to Column and type the starting position for the redefined attribute.

7. To define the attribute's properties, follow steps 6 through 10 for adding a new attribute.

8. Click OK.

   Redefined attributes are listed below other attributes in the properties dialog box.

9. Click Apply to save your changes.

10. Click OK.

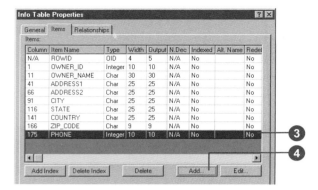

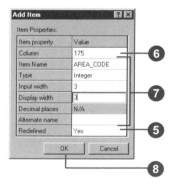

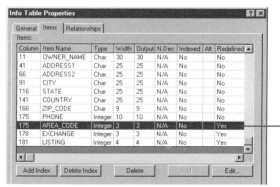

Redefined attributes are added to the bottom of the list.

Tip

Applying changes

When working in the Properties dialog box for a coverage feature class or an INFO table, you might perform a series of changes such as adding or modifying several attributes. At any time, you can click Apply to save your changes.

Modifying an attribute

1. Right-click the coverage feature class or INFO table containing the attributes you want to modify and click Properties.

2. Click the Items tab.

3. Under Item Name, click the attribute you want to modify.

4. Click Edit.

5. In the Value column, click next to the property you want to change. Type a new value or click a new value in the dropdown list, as appropriate. You can only change the Input Width of a redefined attribute.

6. Repeat step 5 until all properties of the attribute have correct values.

7. Click OK.

   The attribute's properties are updated in the properties dialog box.

8. Click Apply to save your changes.

9. Click OK.

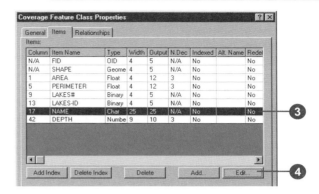

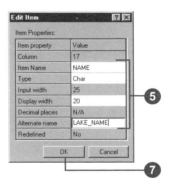

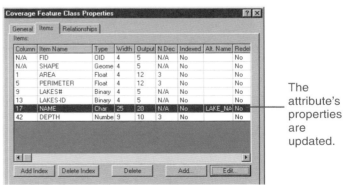

The attribute's properties are updated.

ArcInfo and ArcEditor

Deleting an attribute

1. Right-click the coverage feature class or INFO table containing the attribute you want to delete and click Properties.

2. Click the Items tab.

3. Under Item Name, click the attribute you want to delete.

4. Click Delete.

   The attribute no longer appears in the properties dialog box.

5. Click Apply to save your changes.

6. Click OK.

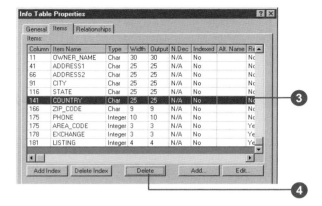

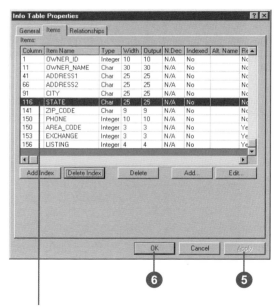

The attribute no longer appears in the properties dialog box.

Adding and removing indexes

1. Right-click the coverage feature class or INFO table whose indexes you want to modify and click Properties.

2. Click the Items tab.

3. Under Item Name, click the attribute to which you want to add or from which you want to remove an index.

4. Click Add Index to create an index of the attribute's values.

   Click Delete Index to remove an existing index on the attribute's values.

   The value in the Indexed column changes to Yes or No depending on whether an index was added or deleted.

5. Click Apply to save your changes.

6. Click OK.

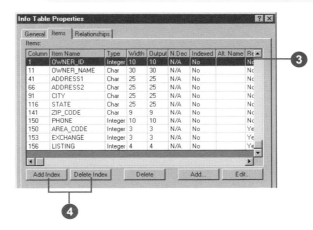

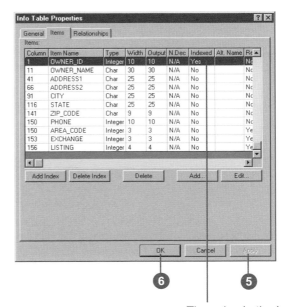

The value in the Indexed column changes to Yes or No depending on whether an index was added or deleted.

ArcInfo and ArcEditor

What is a relationship class?

Features in a coverage often have associations with features in other coverages or attributes in separate tables. You can define these associations in the Catalog by creating relationships, which are stored in *relationship classes*. On the surface, a relationship is similar to an ArcInfo *relate*, although relationships let you represent associations more accurately.

With a relationship, you can define which column in a feature class's attribute table and which column in another table share the same values. Once created, the relationship lets you establish a temporary connection between a coverage's features and descriptive attributes in a table. You can use the related attributes to label, symbolize, or query the features; you can also edit them when editing the features in ArcMap.

Properties of a relationship

One property of a relationship is its *cardinality*, which describes how many features in the coverage are related to how many records in the other attribute table. If the associated table contains measurements taken at a point in the coverage, the relationship will be one to many: one point to many measurements. In general, relationships can have one-to-one (1-1), one-to-many (1-M), many-to-one (M-1), and many-to-many (N-M) cardinalities.

In the example above, the point feature class in the coverage is the *origin* of the relationship, and the table containing the measurements is the *destination*. The columns used to connect these data sources are key attributes. The point feature class, the origin, has an attribute containing a code for each station; this is the *primary key* for the relationship. The measurements table has an attribute indicating which station the measurements were taken at; this is an embedded *foreign key*.

Relationships have path labels that describe the nature of the association. The forward path label describes the relationship

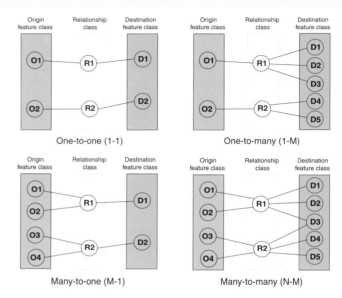

One-to-one (1-1) One-to-many (1-M)

Many-to-one (M-1) Many-to-many (N-M)

Relationships have cardinality. Cardinality describes how many objects of type A are associated with objects of type B. Relationships can have 1-1, 1-M, M-1, or N-M cardinality.

when navigated from the origin to the destination; for example, station points "have" measurements. The backward path label describes the same relationship when navigated from the destination to the origin; in this example, measurements "are taken at" stations.

Types of relationships

There are two types of relationships: simple and composite. *Simple relationships* describe associations between data sources that exist independently of each other. A coverage and table are independent of each other if, when you delete the origin coverage, the destination table continues to exist. ▶

In the previous example, if you started taking measurements at a new point upstream and deleted the old point from the coverage, you would still keep the measurements taken from the old station for historical purposes. The relationship, therefore, is a simple one.

Composite relationships describe associations where the lifetime of one object controls the lifetime of its related objects. An example is the association between highways and points for placing a highway shield marker. The primary key in the line feature class in the highways coverage has a unique code for each line. The foreign key in the point feature class in the shields coverage contains the code for the line it is associated with. Shield points can't exist without a highway.

After creating the composite relationship, when you edit highway lines in ArcMap your changes will affect marker points in the shields coverage. When you move, rotate, or delete a highway line, a message is sent to the related points, which are then moved or deleted appropriately. The default is for messages to be sent along the forward path in the relationship, though you can choose to send them backwards as well.

Simple relationships can have one-to-one, one-to-many, or many-to-many cardinalities. A many-to-one relationship is, by definition, a one-to-one relationship. Composite relationships always have a one-to-many cardinality. When you create a one-to-many relationship, whether simple or composite, the "one" side of the relationship must be the origin. The "many" side must be the destination.

One object can participate in many relationships. For example, in addition to the composite relationship between highways and shield points, the highway line feature class might have a simple relationship to an INFO table. In this case, each highway line has a code indicating the type of surface used, and the related table contains a description of each surface code, so many highways share the same surface description. To describe this second association, you would create a simple, one-to-one relationship.

Although all the above examples have included coverages, it is important to note that you can create a relationship class to define an association between two INFO tables.

Coverage relationship classes

As described in Chapter 11, 'Working with maps and layers', layers may define joins and relates between geographic and tabular data that are stored in different formats or in different ArcInfo workspaces. Joins and relates provide similar functionality to simple relationship classes, except that they must be defined for each individual layer.

You can create a coverage relationship class to model the relationships between objects in an ArcInfo workspace in a permanent and realistic fashion. Once created, the information can be reused in many layers. ArcMap will detect when a relationship class exists and let you easily access and edit related attributes. With composite relationships, as you edit features in a coverage, related features and attributes are edited appropriately.

Coverage relationship classes are essentially the same as the relationship classes you can create in a geodatabase. For specific information regarding relationship classes in geodatabases, see *Building a Geodatabase* or the appropriate topics in the online Help system.

ArcInfo and ArcEditor

Creating a coverage relationship class

You can create a simple or composite relationship class between any two tables and feature classes in the same folder that share a common attribute. Related feature classes may exist in the same coverage or in different coverages. Create a simple relationship if the objects in the origin and destination exist independently of each other. Create a composite relationship if the lifetime of the objects in the origin controls the lifetime of the objects in the destination.

A relationship class lets you query, label, and symbolize the features in the coverage using attributes in the associated table. With any relationship class, you can edit attribute values in the destination in ArcMap while editing the origin coverage's features. For example, with a composite relationship class, when you move, rotate, or delete the power lines in the origin coverage, ArcMap automatically moves or deletes poles in the destination coverage.

The relationship classes in which a coverage feature class or an INFO table participates are listed in the Relationships ▶

1. In the Catalog tree, click the folder containing the coverage for which you want to create a relationship class.

2. Click File, point to New, then click Coverage Relationship Class.

3. Type a name for the new relationship class.

4. Click the origin table or feature class.

5. Click the destination table or feature class.

6. Click Next.

7. Click the type of relationship you want to create.

8. Click Next. ▶

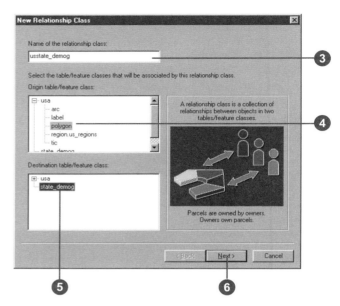

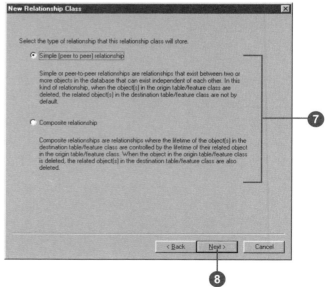

tab in their Properties dialog box. Click a specific relationship in the list and click Properties to learn more about it. For example, you can determine all the coverages and tables to which an INFO table is related. An item's relationships are also recorded in its metadata; with the ESRI stylesheet, you'll find the item's relationships listed at the bottom of the Attributes tab in the metadata.

9. Type the forward and backward path labels for the relationship.

10. Click Next.

11. Click the appropriate cardinality for this relationship.

12. Click Next. ►

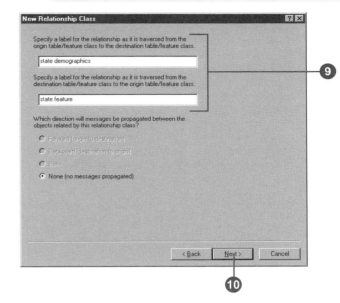

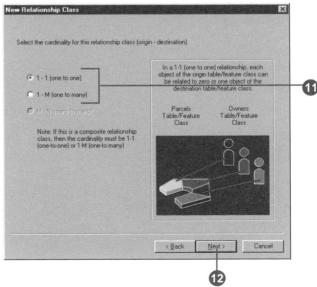

13. Click the first dropdown arrow to see a list of attributes in the origin table or feature class. Click the primary key for this relationship.

14. Click the second dropdown arrow to see a list of attributes in the destination table or feature class. Only those attributes that have the same data type as the primary key are listed. Click the foreign key for this relationship.

15. Click Next.

16. Review the options you specified for the new relationship class. If you want to change something, you can go back through the wizard by clicking the Back button.

17. When satisfied with your options, click Finish.

    The new relationship class appears in the folder's contents.

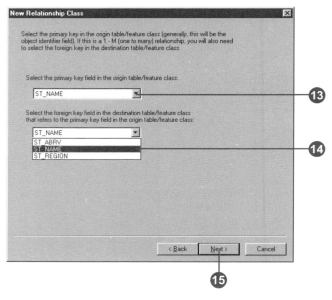

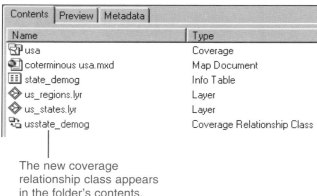

The new coverage relationship class appears in the folder's contents.

Geocoding addresses

14

When you want to map the locations of addresses, you need to create spatial descriptions of these locations from the textual descriptions contained in the addresses. ArcGIS uses *geocoding services* to perform the task of finding the locations of addresses.

You can use ArcCatalog to create geocoding services for finding the locations of addresses. You can also create new feature classes containing the locations of addresses in ArcCatalog by geocoding tables containing address information. This chapter describes the key concepts and tasks associated with creating and managing your geocoding services and geocoding tables of addresses.

Geocoding services in ArcCatalog

What is a geocoding service?

Geocoding is also commonly known as address matching. It is the process of creating a spatial description of a place, such as a point feature, from a nonspatial descripton of that place such as a street address. A geocoding service defines a process for converting nonspatial descriptions of places into spatial descriptions.

A geocoding service defines paths to reference data, rules for standardizing alphanumeric descriptions of places and matching them to the reference data, and parameters for reading address data and creating output. Geocoding services are based on geocoding service styles. Several geocoding service styles are provided that you can use to create geocoding services.

Finding geocoding services in ArcCatalog

You can create and use both clientside and serverside geocoding services in ArcCatalog and ArcMap. Clientside geocoding services are stored on your computer, and serverside, or ArcSDE, geocoding services are stored in an ArcSDE geodatabase. To create a new geocoding service, double-click Create New Geocoding Service in the appropriate Geocoding Services folder.

Clientside geocoding services are listed within the Geocoding Services folder located immediately under the Catalog in the Catalog tree. By default, new clientside geocoding services are private; they aren't shared with other users who might log in to the same computer. You can make a geocoding service public, which allows other users on the same computer to use it. You can distinguish between public and private geocoding services by their names; private services are prefixed by your username, for example, "jon.StreetAddresses".

Each ArcSDE database connection also has a Geocoding Services folder that lists the geocoding services that are stored within that

geodatabase. Serverside geocoding services are easily shared among users in an organization. Although you may see a geocoding service listed in an ArcSDE Geocoding Services folder, you will be able to use it only if you can access the reference data on which the geocoding service is based. For more information, see 'Managing geocoding services'.

In ArcCatalog you can also see and use geocoding indexes created with ArcView GIS 3. These are stored in the same folder as the shapefile or coverage on which they are based. ArcView GIS 3 geocoding indexes based on shapefiles have the same name as the corresponding shapefile with an .mxs file extension. ArcView GIS 3 geocoding indexes based on coverages have the same name as the coverage, usually with an .mxa file extension.

Geocoding indexes

To create a new geocoding service, double-click Create New Geocoding Service in the appropriate Geocoding Services folder. When a new ArcGIS geocoding service is created, a geocoding index is built so that features in the reference data can be quickly matched to the addresses that you want to geocode.

The information contained in the geocoding indexes is determined by the style on which the geocoding service is based. The geocoding service styles provided with ArcGIS software let you build geocoding indexes on fields that contain street name information, both for the primary reference data feature class and the alternate street name table, if any. Geocoding services styles that use shapefiles or coverages as reference data also create indexes for house number ranges and zones, if applicable to the particular style.

Geocoding indexes are stored as tables in a geodatabase or as files on disk along with the reference data. However, they aren't visible in ArcCatalog.

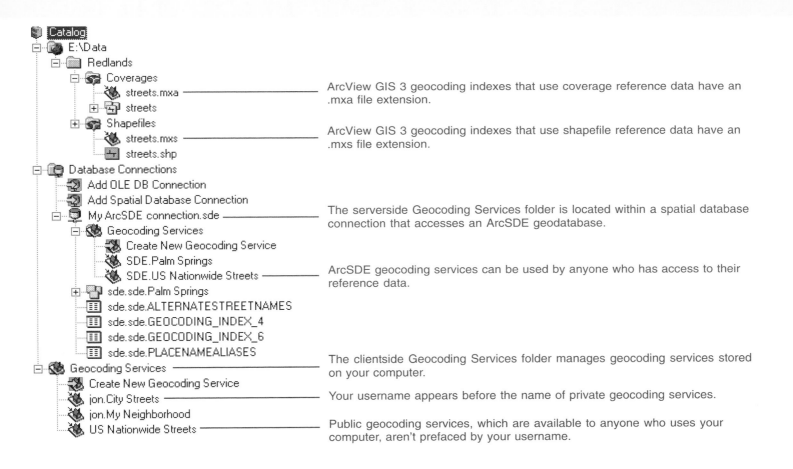

Catalog
- E:\Data
  - Redlands
    - Coverages
      - streets.mxa ———————— ArcView GIS 3 geocoding indexes that use coverage reference data have an .mxa file extension.
      - streets
    - Shapefiles
      - streets.mxs ———————— ArcView GIS 3 geocoding indexes that use shapefile reference data have an .mxs file extension.
      - streets.shp
- Database Connections
  - Add OLE DB Connection
  - Add Spatial Database Connection
  - My ArcSDE connection.sde ———————— The serverside Geocoding Services folder is located within a spatial database connection that accesses an ArcSDE geodatabase.
    - Geocoding Services
      - Create New Geocoding Service
      - SDE.Palm Springs
      - SDE.US Nationwide Streets ———————— ArcSDE geocoding services can be used by anyone who has access to their reference data.
    - sde.sde.Palm Springs
    - sde.sde.ALTERNATESTREETNAMES
    - sde.sde.GEOCODING_INDEX_4
    - sde.sde.GEOCODING_INDEX_6
    - sde.sde.PLACENAMEALIASES
- Geocoding Services ———————— The clientside Geocoding Services folder manages geocoding services stored on your computer.
  - Create New Geocoding Service
  - jon.City Streets ———————— Your username appears before the name of private geocoding services.
  - jon.My Neighborhood
  - US Nationwide Streets ———————— Public geocoding services, which are available to anyone who uses your computer, aren't prefaced by your username.

Managing geocoding services

You can create and manage geocoding services in ArcCatalog. When you create a new geocoding service, you specify the data that the geocoding service will use to determine the locations of addresses. In addition to using feature classes, shapefiles, or coverages with address attributes, you can optionally define alternate street name tables and place name alias tables to use during the geocoding process. However, before you can create a geocoding service, you must prepare the reference data that the geocoding service will use. After creating a new geocoding service, you may wish to make it public so that others can use it as well.

See Also

For information on preparing reference data for a geocoding service and the available geocoding service styles, see the topic 'Preparing reference data for a geocoding service' in the 'Working with geodatabases' section of the online Help.

Creating a new geocoding service

1. In ArcCatalog, click a Geocoding Services folder.

2. Double-click the Create New Geocoding Service item.

3. Click the geocoding service style that you want to use to create the new geocoding service.

4. Click OK. ▶

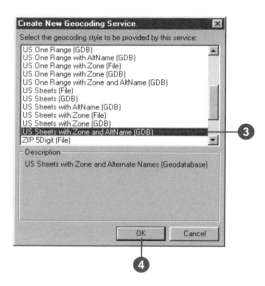

5. In the Name text box, type a name for the new geocoding service.

6. Click the Browse button on the Primary Table tab.

7. Navigate to and click the feature class, shapefile, or coverage that the geocoding service will use as reference data, then click Add.

8. Click a dropdown arrow, then click the name of the column that contains the specified address information.

The required address attributes have a bold name to the left of the appropriate dropdown list. ▶

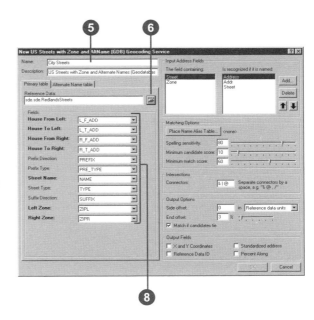

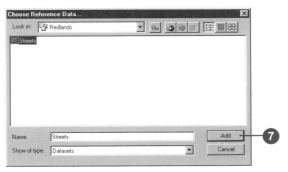

9. If your geocoding service will use an alternate street name table, click the Alternate Name table tab.

Otherwise, skip to step 13.

10. Click the Browse button.

11. Navigate to and click the table that the geocoding service will use as an alternate street name table, then click Add.

12. Click a dropdown arrow, then click the name of the column that contains the specified alternate street name information.

The required alternate street name attributes have a bold name to the left of the appropriate dropdown list.

13. If your geocoding service will use a place name alias table, click Place Name Alias Table.

Otherwise, skip to step 16. ▶

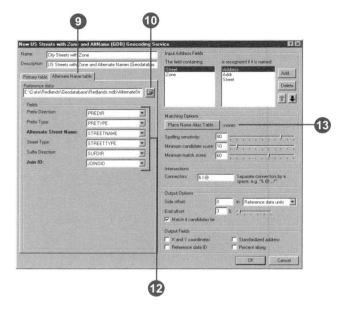

14. Click the Browse button, navigate to and click the table that the geocoding service will use as a place name alias table, then click Add.

15. Click a dropdown arrow, then click the name of the column that contains the specified place name alias information.

    The required place name alias attributes have a bold name to the left of the appropriate dropdown list.

16. Click OK.

17. Click OK to create the new geocoding service.

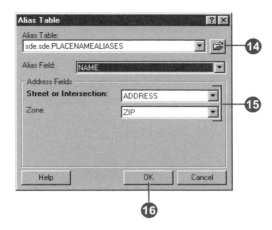

ArcInfo and ArcEditor

Making a serverside geocoding service public

1. Click the ArcSDE database connection that contains the geocoding service that you want to make public.

2. Right-click the feature dataset that contains the feature class, the standalone feature class, or the table that the geocoding service uses as reference data and click Privileges.

3. Type the name of the user to which you want to make the geocoding service available.

4. Check SELECT.

5. Click Apply.

6. Repeat steps 3 through 5 to make your geocoding service available to additional users.

7. Click OK.

8. Repeat steps 2 through 7 for the alternate street name tables and place name alias tables if your geocoding service uses them.

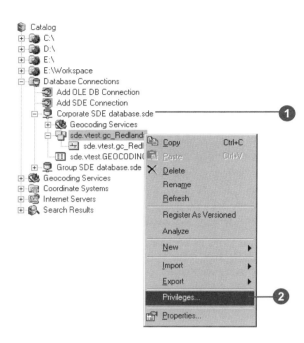

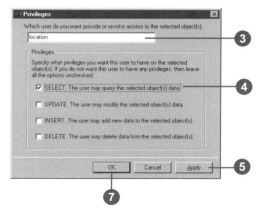

Controlling the geocoding process

The geocoding process

Once you have created a geocoding service, you can begin right away to geocode addresses with it. However, understanding how a geocoding service matches addresses and how modifying a geocoding service's settings affects this process can help you improve both the performance and accuracy of your geocoding.

Geocoding services use a specific process to match an address. First, the geocoding service standardizes the address. Second, the geocoding service searches the reference data to find potential candidates. Next, each potential candidate is assigned a score based on how closely it matches the address. Finally, the address is matched to the candidate with the best score.

When a geocoding service standardizes an address, it dissects the address into its address components. For example, the address "4 Dundas St. E." has four address components: the street number, "4"; the street name, "Dundas"; the street type, "St."; and the street direction, "E." Each style of geocoding service standardizes an address into a different set of address components.

A geocoding service standardizes an address into a number of address components.

If your geocoding service uses a place name alias table, it searches this table for entries that match the address you are trying to geocode to determine if the address is actually a place name alias. If one is found, it substitutes the address in the place name alias table for the place name that you are trying to locate and standardizes this address.

Once it has standardized the address, the geocoding service searches the reference data to find features with address components that are similar to the components of the standardized address. Each style of geocoding service bases this search on a different set of address components. The geocoding service uses its spelling sensitivity setting for some address components, such as street name, to determine how closely the address components of a feature must match the address components of the address you are geocoding. If the geocoding service uses an alternate street name table, then it also searches this table to find potential candidates.

Street Name	Street Type	Street Direction
Dundas	St	E
Dundas	St	W
Centre	St	
Richmond	Blvd	E
Dundalk	Ct	
Dunrobin	St	

The geocoding service searches the reference data for features with address components that are similar to the components of the standardized address.

Once the geocoding service has generated a set of potential candidates, it scores each potential candidate in order to determine how closely each potential candidate matches the address that you are geocoding. Each potential candidate is assigned a score from 0 to 100. Each address component is used to generate this score. The score for each potential candidate will be lower if address components are misspelled (for example, the street name is misspelled), incorrect (for example, the street number of the address does not fall within the address range for the candidate), or missing (for example, if the street direction is specified in the address but not in the potential candidate). Once each potential candidate is scored, the geocoding service generates a set of candidates that are potential matches for the address. Which ▶

potential candidates are considered to be candidates is based on the geocoding service's minimum candidate score setting.

From Address	To Address	Street Name	Street Type	Street Direction	Match Score
1	99	Dundas	St	E	100
1	99	Dundas	St	W	75
1	49	Dundalk	Ct		10
1	99	Dunrobin	St		25

The geocoding service scores each potential candidate using all of the address components. A set of match candidates is generated based on the geocoding service's minimum candidate score setting.

Finally, the geocoding service finds the candidates with the highest score. If the score of the candidate with the best score exceeds the geocoding service's minimum match score setting, then the geocoding service matches the address to that candidate.

Geocoding service settings

Geocoding services have a number of settings that you can use to control the geocoding process. These settings control how a geocoding service reads a table of addresses that you want to geocode, how it matches addresses to features in the reference data, and what it writes to the geocoded output. Modifying a geocoding service's settings will impact how well the geocoding service will be able to match addresses to the reference data, as well as what information the geocoding result will contain.

Input address fields

When you geocode a table of addresses, a geocoding service will try to determine which fields in the table contain the address information. A geocoding service accomplishes this by searching for default input address field names that you define for the geocoding service. When you geocode a table of addresses and the geocoding service finds a field with one of the default

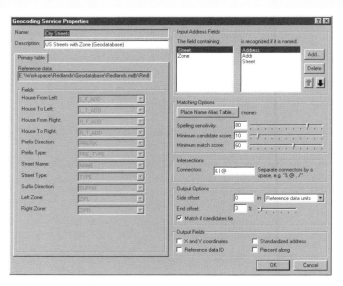

The Geocoding Service Properties dialog box allows you to modify the settings of a geocoding service.

address field names that you defined, then the geocoding service will automatically use the contents of this field for a particular address component.

You can use the input address field settings of a geocoding service to specify names of fields in your address tables that are likely to contain address information. For example, a field that contains the zone component of an address might be called "City", "ZIP", "ZIPCode", or "Zone". If your geocoding service requires zone information for addresses that it can geocode, then you can use the input address field settings to make the geocoding service search for fields with these names when it tries to find the zone information in a table of addresses.

If the geocoding service cannot find a field in the address table with one of the default names that you specified, then you will ▶

have to manually choose which fields contain the address information. For more information on the input address components that each style of geocoding service requires, see the online Help.

Matching options

Spelling sensitivity

The spelling sensitivity setting controls how much variation the geocoding service will allow when it searches for likely candidates in the reference data. A low value for spelling sensitivity will allow "Mane", "Maine", and "Man" to be treated as match candidates for "Main". A higher value will restrict candidates to exact matches. The spelling sensitivity does not affect the match score of each candidate; it only controls how many candidates the geocoding service considers. The geocoding service then computes the match score of each candidate and ranks the candidates by score.

The spelling sensitivity setting for a geocoding service is a value between 0 and 100. By default, the spelling sensitivity is 80, which does not allow for much variation in spelling. If you are sure that your addresses are spelled correctly, you could set a higher spelling sensitivity. If you think that your addresses may contain spelling errors, then you should use a lower setting. The geocoding process takes longer when you use a lower setting because the geocoding service has to compute scores for more candidates.

Minimum candidate score

When a geocoding service searches for likely candidates in the reference data, it uses this threshold to determine whether a potential candidate should be considered. Candidates that yield a match score lower than this threshold will not be considered.

The minimum candidate score for a geocoding service is a value between 0 and 100. By default, this is set to 30. If the geocoding service seems unable to come up with any likely candidates for an address that you want to geocode, you could lower this setting so that candidates with very low scores are considered.

Minimum match score

The minimum match score setting lets you control how well addresses have to match their most likely candidate in the reference data in order to be considered matched. A perfect match yields a score of 100. A match score between 80 and 100 can generally be considered a good match. An address below the minimum match score is considered to have no match.

The minimum match score for a geocoding service is a value between 0 and 100. By default, this setting is 60. If your application demands that addresses be located with a high level of confidence, you should set a higher minimum match score. If you want to maximize the number of addresses that can be matched and don't mind if some addresses are potentially matched incorrectly, you can use a lower setting.

Intersection connectors

Geocoding services that are based on the US One Range, US Streets, and ArcView StreetMap™ geocoding service styles can geocode street intersections in addition to street addresses. In ArcGIS, intersections are designated as two streets delimited by an intersection connector string. Some examples of intersection descriptions are "Hollywood Blvd. & Vine St." and "Yonge and Bloor".

The intersection connectors setting lets you specify all of the strings that the geocoding service will recognize as intersection connectors. By default, "&", "|", and "@" are recognized as intersection connectors. ▶

Output options

Side offset

Some styles of geocoding service use reference data that contains address range information for each side of the street including the US Streets and StreetMap geocoding service styles. Geocoding services based on these styles can determine on which side of the street an address is located. For cartographic purposes, you can specify a side offset for geocoded features when using these styles of geocoding services. When you specify a side offset, the geocoding service locates geocoded features at the specified distance from the street centerline on the correct side of the street.

End offset

Geocoding services that use reference data with line geometry, such as those based on the US One Range, US Streets, or StreetMap geocoding service styles, can interpolate a position along reference features for a geocoded address. In order to prevent features that are located at the end of a reference feature from falling on top of other features (for example, a cross street), the geocoding service can apply a "squeeze factor", or end offset, to the location of a geocoded address. The end offset setting of a geocoding service is expressed as a percentage of the length of the reference feature, between 0 percent and 50 percent. An end offset setting of 0 percent will not offset features from the end of the reference feature. An end offset of 50 percent will locate all features at the middle of the reference feature. By default, the end offset setting for a geocoding service is 3 percent.

Match if candidates tie

If a geocoding service finds two or more reference features that have the same best match score, you can specify whether or not to match an address arbitrarily to one of these features. Use this setting to specify whether to arbitrarily match these addresses or

to leave them unmatched. In either case, you can review addresses with tied candidates during the interactive review process whether or not they are matched. ▶

The address, 100 MAIN ST, has been offset from the street feature by the offset distance of 25 feet. This address falls at the end of the street feature and is therefore inline with the end of the street feature.

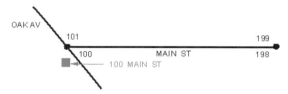

When streets intersect at odd angles, specifying an offset distance can have the undesirable effect of placing the address so it appears that the address does not belong to MAIN ST, but rather to OAK AV.

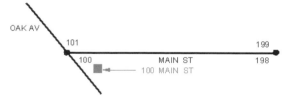

An end offset can be specified that adjusts the location of the address toward the center of the street feature. Using an end offset will often rectify the condition shown in the previous diagram. In this example, a squeeze factor of 10 percent was used to move the address toward the center of the street feature by a distance equal to 10 percent of the length of the street feature.

Output fields

x,y coordinates

Use this setting to specify whether or not to create attributes in geocoded feature classes that contain the x,y coordinates of the geocoded features. If you use this setting, then a geocoding service will create two attributes in the output features classes that you create with it, one each for the x,y coordinates of the geocoded features. These attributes are not valid for an address if the address is not matched.

Standardized addresses

Use this setting to specify whether or not to create an attribute in a geocoded feature class that contains the standardized address. The contents of this field for each address are the address components used by the geocoding service, separated by the pipe ("|") character. This attribute is useful for determining how the geocoding service standardized the addresses.

Reference data ID

Use this setting to specify whether or not to create an attribute in a geocoded feature class that contains the ID of the reference feature to which an address was matched. This attribute is not valid for an address if the address is not matched.

Percent along

Geocoding services based on the US One Range, US Streets, or StreetMap geocoding services styles can interpolate a position along reference features for a geocoded address. Use this setting to specify whether or not to create an attribute in a geocoded feature class that contains the position along the reference feature to which the address was matched. The value of this attribute is a number between 0 and 100, with 0 indicating the starting node of the reference feature and 100 indicating the

ending node of the reference feature. This attribute is not valid for an address if the address is not matched.

Modifying a geocoding service's settings

You can modify a geocoding service's settings to control how it determines the locations of addresses and what information is contained in the geocoding output. One option is to set the columns in which you expect to find street names, for example. When the geocoding service searches an address table for the column containing street names, it looks for columns in the order in which they appear in the right-hand side list under Input Address Fields. If the table doesn't contain a column named "Address" it will look for a column named "Addr". You might also set the sensitivity to use when matching addresses, or define which attributes should be added to the resulting feature class.

Tip

Reordering default input column names

To change the order in which the column names appear, click a name, then click the up and down arrows to move the name up or down in the list.

Adding a default input address column name

1. Right-click the geocoding service that you want to modify and click Properties.

2. In the list on the left-hand side under Input Address Fields, click the input address column for which you want to search in address tables.

3. Click Add.

4. Type the name of the field to search for in address tables, then click OK.

5. Click OK.

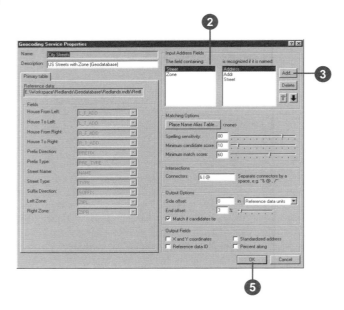

Setting matching options

1. Right-click the geocoding service that you want to modify and click Properties.

2. Under Matching Options, drag the spelling sensitivity slider to the desired setting.

3. Drag the minimum candidate score slider to the desired setting.

4. Drag the minimum match score slider to the desired setting.

5. Click OK.

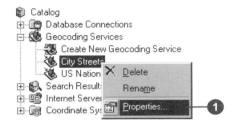

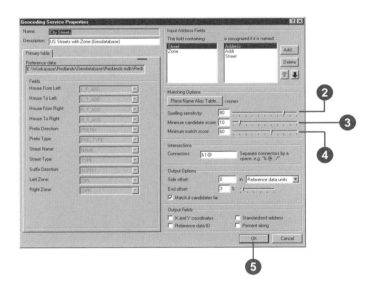

Specifying intersection connectors

1. Right-click the geocoding service that you want to modify and click Properties.

2. Type the intersection connectors that the geocoding service will recognize in the Intersection Connectors text box.

    Items in the Intersection Connectors text box must be separated by a space (for example, "& @ , |").

3. Click OK.

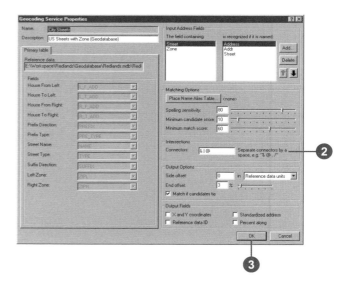

Setting output options

1. Right-click the geocoding service that you want to modify and click Properties.

2. Under Output Options, click the dropdown arrow and click the units that will be used to measure the side offset.

3. Type the number of units to offset geocoded addresses.

4. Drag the end offset slider bar to the desired setting.

5. Check Match if candidates tie to match addresses arbitrarily when two or more candidates with the same best match score exist.

6. Click OK.

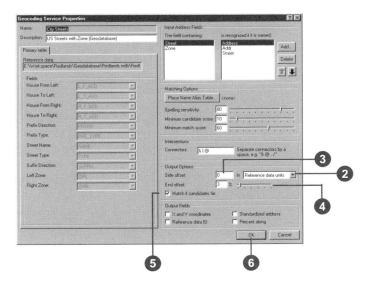

Specifying output attributes

1. Right-click the geocoding service that you want to modify and click Properties.

2. Under Output Fields, check X and Y Coordinates to write X and Y coordinates of geocoded features to geocoded feature classes.

3. Check Standardized address to write standardized addresses to geocoded feature classes.

4. Check Reference data ID to write the IDs of the reference data features to which addresses were matched to geocoded feature classes.

5. Check Percent Along to write the percent along reference features at which addresses are located to geocoded feature classes.

6. Click OK.

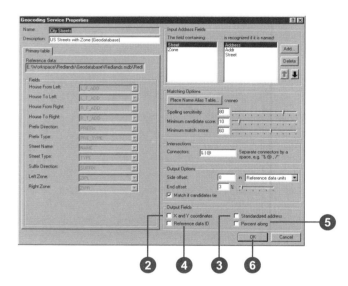

Geocoding a table of addresses

When you geocode a table of addresses, you use a geocoding service to create point features that represent the locations of the addresses. The output point features can be stored in any data format supported by ArcCatalog including geodatabase feature classes and shapefiles; ArcView users can't store output features within an ArcSDE geodatabase. However, before you can geocode a table of addresses, you must create a geocoding service, and you must prepare your table to be geocoded. For information on preparing your address tables for geocoding, see the topic 'Preparing address data for geocoding' in the section 'Working with geodatabases' in the online Help.

1. In ArcCatalog, right-click the table of addresses that you want to geocode and click Geocode Addresses.

2. Click the geocoding service that you want to use to geocode this table of addresses.

   If the geocoding service that you want to use does not appear in this list, click Browse for service, then browse for the geocoding service that you want to use.

3. Click OK ▶

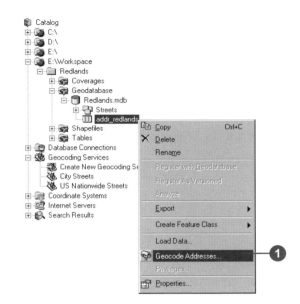

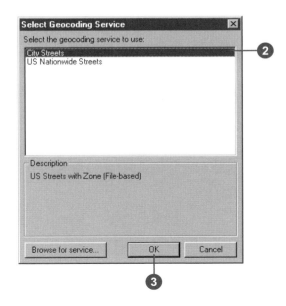

4. If the geocoding service did not automatically find the fields in the address table that contain the address information, click the appropriate dropdown list and click the name of the column that contains the appropriate address attribute.

   The required address attributes have a bold name to the left of the appropriate dropdown list.

5. Click the Browse button.

6. Navigate to and click the location where you want to create the geocoded feature class, then click Save.

7. Click Create a dynamic feature class related to the table if appropriate.

8. Click Advanced Geometry Options to specify the geometry settings for the geocoded feature class. ▶

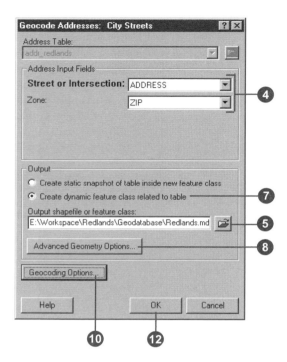

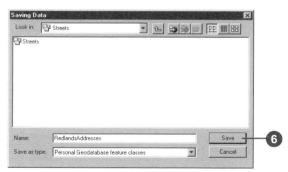

Specifying the geocoding settings

When you are geocoding a table of addresses, modifying the geocoding settings does not change the geocoding service that you are using. Only the settings that are used to geocode this table are modified. These settings are stored with the geocoded feature class. The original geocoding service is not modified.

9. Specify the geometry settings for the geocoded feature class and click OK.

10. Click Geocoding Options to specify the geocoding options that will be used to geocode the table of addresses.

11. Specify the geocoding settings that you want to use to geocode the table of addresses and click OK.

12. Click OK on the Geocode Addresses dialog box to begin geocoding the table. ▶

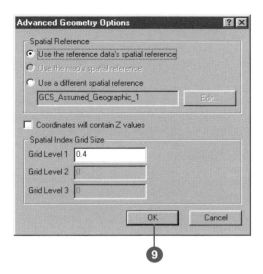

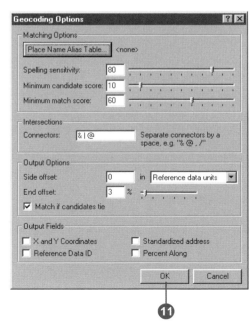

Attributes in geocoded feature classes

When you geocode a table of addresses, a geocoding service creates some special attributes in the output feature class.

The Status attribute indicates whether or not the address was matched. This attribute has values of "M" for matched addresses, "U" for unmatched addresses, and "T" (tied) for addresses for which there were more than one candidate with the best match score.

The Score attribute contains the match score of the candidate to which the address was matched.

The Side attribute contains the side of the street to which an address was matched, if the geocoding service that was used to match the table contains address information for both sides of the street. This attribute has values of "L" for the left side of the street, "R" for the right side of the street, or nothing if the geocoding service could not determine the side of the street.

13. Review the results of the geocoding process.

14. Click Done.

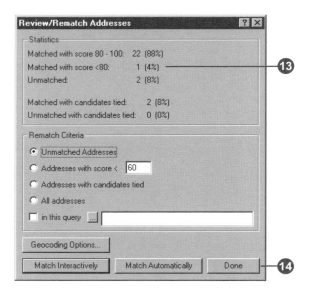

Rematching a geocoded feature class

After you have geocoded a table of addresses, you may want to review the results. If you are unhappy with the results, you may want to modify the geocoding service's settings and try to geocode the the table of addresses again. This process is known as *rematching*. There are a number of options for specifying which addresses in a geocoded feature class you want to rematch. You can rematch just the addresses that are un-matched, all of the addresses with a match score less than a certain value, all of the ad-dresses with two or more candidates with the best match score, or all of the addresses. In addition, you can specify a query to use that defines the set of addresses to rematch.

See Also

For more information on defining queries, see Using ArcMap.

Rematching a geocoded feature class automatically

1. In ArcCatalog, right-click the feature class that you want to rematch and click Review/Rematch Addresses.

2. Specify the criteria for the addresses that you want to rematch.

3. Click Geocoding Options to modify the geocoding settings that you want to use to rematch the addresses. ▶

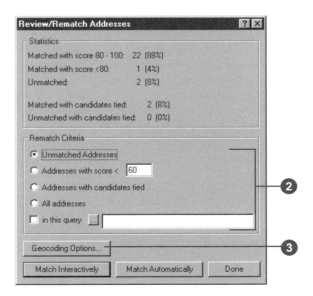

Tip

Specifying the geocoding settings

When you are rematching a geocoded feature class, modifying the geocoding settings does not change the geocoding service that was originally used to create the feature class. The geocoding settings that are presented here are the settings that were used to geocode the feature class. The original geocoding service is not modified.

4. Specify the geocoding settings that you want to use to rematch the geocoded feature class and click OK.

5. Click Match Automatically.

6. Review the results of rematching the specified addresses.

7. Click Done.

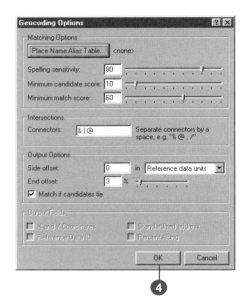

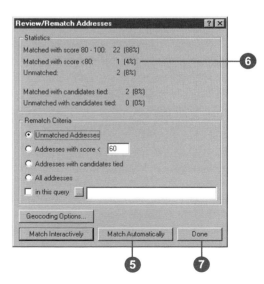

Rematching a geocoded feature class interactively

1. In ArcCatalog, right-click the feature class that you want to rematch and click Review/Rematch Addresses.

2. Specify the criteria for the addresses that you want to rematch.

3. Click Match Interactively. ▶

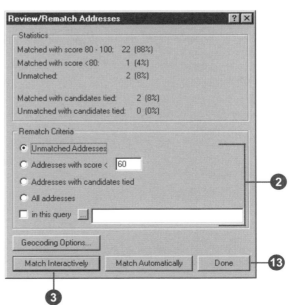

Tip

Unmatching an address

Sometimes you may want to unmatch an address. For example, the geocoding service may have matched an address to a candidate, but you may not be happy with any of the candidates for the address. In the list of candidates, click the candidate to which the address is matched and click Unmatch.

4. Click the address in the top list that you want to rematch.

5. If necessary, edit the input address.

6. Check the address standardization to ensure that the geocoding service has standardized the address correctly. If not, edit the address standardization by clicking Modify.

7. Click Geocoding Options to modify the gecoding settings that you want to use to rematch the addresses.

8. Specify the geocoding settings that you want to use to geocode the addresses and click OK.

9. Click Search to refresh the list of candidates.

10. Click the candidate in the bottom list to which you want to match the address.

11. Click Match.

12. Click Close when you are finished rematching the addresses.

13. Click Done to finish rematching the geocoded feature class.

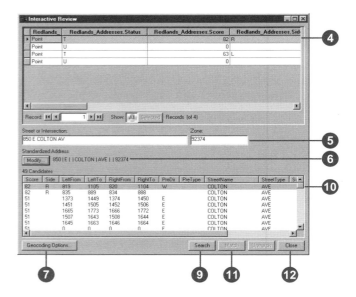

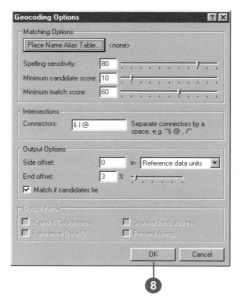

Customizing ArcCatalog 15

After working with ArcCatalog a while, you might want to customize its appearance to reflect your own preferences. Whether you want to hide toolbars you don't use, modify toolbars and their contents, or create a custom toolbar, you can do it without writing a single line of code. You can add custom commands that others have created or, if you know Visual Basic, you can write macros and create custom commands yourself. This chapter will help you start tailoring the Catalog's user interface to meet your needs. More detailed information can be found in *Exploring ArcObjects*.

Basic user interface elements

ArcCatalog has Main Menu and Standard toolbars. Both are referred to as *toolbars*, although the Main Menu toolbar contains menus only.

Any toolbar can be *docked* at the top or bottom or to the left or right side of the ArcCatalog window. Alternatively, toolbars can float on the desktop while functioning as part of the application. When a toolbar is docked, it is moved and resized with the ArcCatalog window. The Catalog tree is docked on the left by default, but you can dock it elsewhere in the window or have it float on the desktop. To prevent a toolbar or the Catalog tree from docking, hold down the Ctrl key while dragging it.

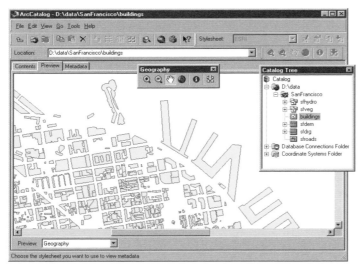

The ArcCatalog window with the Geography toolbar and the Catalog tree floating on the desktop. All other toolbars are docked.

Toolbars can contain menus, menu items, buttons, tools, combo boxes, and edit boxes; these are different types of *commands*. Each type of command lets you interact with the user interface differently.

- *Menus* arrange other commands into a list.
- *Buttons* and *menu items* perform actions when clicked.
- *Tools* require interaction with the user interface before an action is performed. The Zoom In command is a good example of a tool—when you click or drag a rectangle over data in Geography view, ArcMap redraws the data at a larger scale.
- *Combo boxes* let you type information or choose an option from a dropdown list. For example, the Location combo box on the Standard toolbar lets you select an item by typing its path or by choosing a path from its dropdown list.
- *Text boxes*, or *edit boxes*, let you type in text. In ArcMap, you can type the scale at which you want to view the map.

Each command is associated with code. When you click a command, the code for the click event starts running. *Events* are fired when you interact with a command. Each type of command behaves differently because it has a different set of properties, methods, and events.

Changing the Catalog's appearance

You can group commands, add new macros, or add custom commands in ArcCatalog using the Customize dialog box. The Customize dialog box resembles and has many of the same properties as the equivalent dialog box in Microsoft Office 2000 applications. If you've customized any of those applications, the process will be familiar to you.

The Customize dialog box lets you modify existing menus, toolbars, and context menus with simple drag-and-drop techniques. Afterwards, if you prefer, you can return the menus and toolbars that are built into the Catalog to their default settings. You can create your own menus and toolbars, too.

Customizing toolbars

All of the Catalog's toolbars are visible by default; if a toolbar's commands don't work with the current view, the toolbar is unavailable. To change which toolbars are available, use the Toolbars list in the View menu or the Customize dialog box. A check mark next to the toolbar's name indicates that it's available. Although it appears in the list, you can't hide the Main Menu. After showing a toolbar for the first time, it appears floating on the desktop. If it was previously turned on, it returns to its last position. You can resize floating toolbars to see their commands however you like. A toolbar remembers its floating size even after you hide or dock it. To quickly hide a floating toolbar, click its Close button.

Customizing toolbars lets you tailor ArcCatalog into a powerful and efficient application. For example, you can create your own toolbars to quickly access your most frequently used ArcToolbox tools. You can rename and delete custom toolbars, but you can't rename or delete toolbars that are provided by default with, or built into, ArcCatalog. If a toolbar comes from an ActiveX® dynamic link ▶

Hiding and showing toolbars from the View menu

1. Click the View menu and point to Toolbars.

2. Check a toolbar to show it.

   Uncheck a toolbar to hide it.

Hiding and showing toolbars from the Customize dialog box

1. Click the Tools menu and click Customize.

2. Click the Toolbars tab.

3. Check a toolbar to show it.

   Uncheck a toolbar to hide it.

4. Click Close.

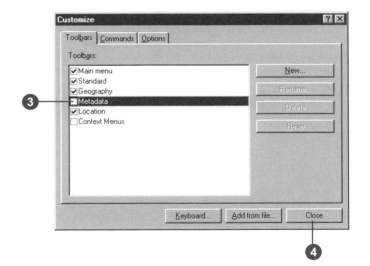

library (DLL) that you added to the Catalog with the Add from file button, it cannot be renamed.

In addition to changing which toolbars appear and adding your own toolbars, you can specify how all toolbars in the Catalog will behave using the Options tab. The commands on a toolbar can appear with large icons rather than the small images you see by default. You can also choose whether or not ToolTips appear when you hold the mouse pointer over a command.

Creating a new toolbar

1. Click the Tools menu and click Customize.

2. Click the Toolbars tab.

3. Click New.

4. Type the name of your new toolbar.

5. Click OK.

   The new, empty toolbar appears in the Toolbars list and as a floating toolbar on your desktop.

6. Click Close.

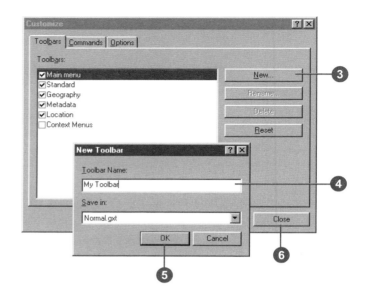

Renaming a toolbar

1. Click the Tools menu and click Customize.

2. Click the Toolbars tab.

3. Click the toolbar you want to rename.

4. Click Rename.

5. Type the name of your new toolbar.

6. Click OK.

   The toolbar is renamed in the Toolbars list, and it appears as a floating toolbar on your desktop.

7. Click Close.

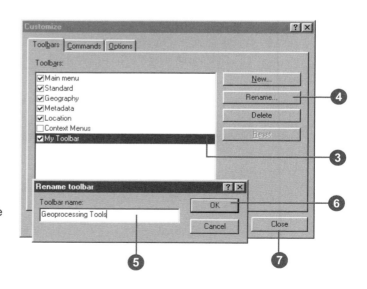

Displaying toolbars with large icons

1. Click the Tools menu and click Customize.

2. Click the Options tab.

3. Check Large icons to display large icons for a toolbar's commands.

4. Click Close.

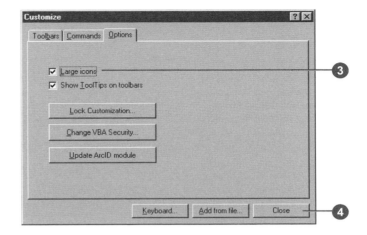

You see large icons on the toolbars.

Showing ToolTips on toolbars

1. Click the Tools menu and click Customize.

2. Click the Options tab.

3. Check Show ToolTips on toolbars to display ToolTips when you hold the mouse pointer over a button.

4. Click Close.

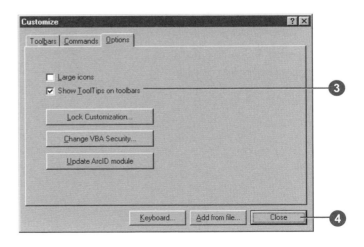

Changing a toolbar's contents

You can modify the contents of any toolbar by adding, moving, and removing commands. Many commands are built into the Catalog. If ArcToolbox is installed on your computer, its tools are also available in the Catalog as commands. You can add any command to any toolbar or menu or add a new menu to a toolbar. You can also modify context menus. Context menus provide easy access to frequently used commands; you see them when you right-click an item in the Catalog. After modifying a built-in toolbar, you can return it to its original contents; you might want to do this if you accidentally remove a command from the toolbar.

Adding a command to a toolbar or menu

1. Show the toolbar to which you want to add a command.

2. Click the Tools menu and click Customize.

3. Click the Commands tab.

4. In the Categories list, click the category that contains the command you want to add.

5. Click the command you want to add in the Commands list.

6. Drag the command to any location in the target toolbar or menu. A menu's contents will appear if you hold the mouse pointer over the menu.

   A black line indicates where the command will be positioned.

7. Drop the command.

   The command appears in the toolbar or menu.

8. Click Close.

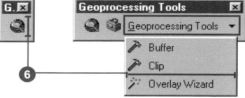

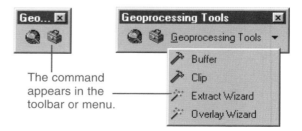

The command appears in the toolbar or menu.

Creating access keys

All menus on the Main Menu and their commands have an underlined character in their caption called an access key. It lets you access the menu from the keyboard by holding down Alt and then pressing the underlined letter. To create an access key, place an ampersand (&) in front of a letter in the menu's (or the command's) caption.

Adding a new, empty menu to a toolbar

1. Show the toolbar to which you want to add a new, empty menu.

2. Click the Tools menu and click Customize.

3. Click the Commands tab.

4. Click New Menu in the Categories list.

5. Click and drag the New Menu command from the Commands list and drop it on the toolbar.

   An empty menu called "New Menu" appears in the toolbar.

6. Right-click New Menu in the toolbar.

7. Type an appropriate caption for the menu in the text box.

8. Press Enter.

9. Click Close.

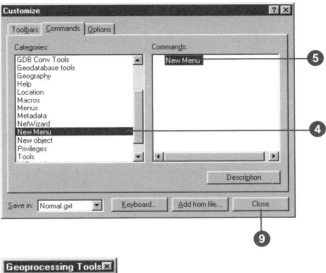

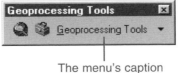

The menu's caption changes.

Adding a command to a context menu

1. Click the Tools menu and click Customize.

2. Click the Toolbars tab.

3. Check the Context Menus toolbar.

4. Click Context Menus on the toolbar.

   A list of all the context menus in the Catalog appears. You may need to scroll up or down in the list to find the menu you want to modify.

5. Click the arrow for the context menu to which you want to add a command.

   The context menu appears.

6. Click the Commands tab in the Customize dialog box.

7. Click the category that contains the command you want to add to the menu.

8. Click and drag the command from the Commands list to the context menu. A thick black line indicates where the command will be positioned. Drop the command in the appropriate location.

   The command appears in the context menu.

9. Click Close in the Customize dialog box.

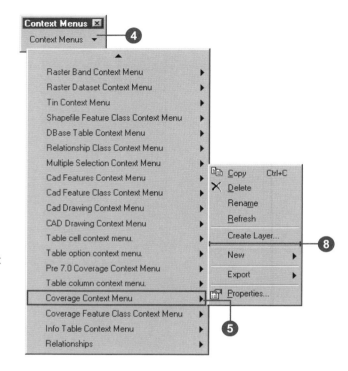

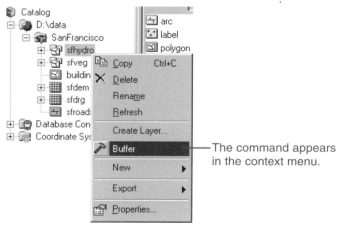

The command appears in the context menu.

Why open the Customize dialog box?

Even though you don't make use of it in an operation such as moving commands, you must display the Customize dialog box when you are customizing the Catalog. With the dialog box open, the Catalog is in a state where you can change its user interface.

Tip

Removing commands

When you remove a command from a toolbar, you're not deleting it; it's merely no longer available on the toolbar. The command still appears in the Commands list in the Customize dialog box. Later, you can always add the command back to the toolbar or reset the toolbar's contents.

Moving a command

1. Show the toolbar with the command you want to move.

2. If you're moving the command to another toolbar, show the destination toolbar.

3. Click the Tools menu and click Customize.

4. Click and drag the command from its original position and drop the command in its new location.

   The command appears in the new position.

5. Click Close in the Customize dialog box.

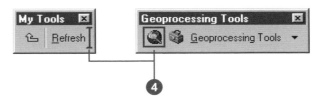

The command appears in its new position.

Removing a command

1. Show the toolbar containing the command that you want to remove.

2. Click the Tools menu and click Customize.

3. Click and drag the tool you want to remove from the toolbar.

   The mouse pointer changes to a line through a circle.

4. Drop the command.

   The command is removed from the toolbar or menu.

5. Click Close in the Customize dialog box.

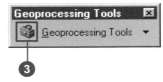

The command is removed from the toolbar.

Resetting a built-in toolbar

1. Click the Tools menu and click Customize.

2. Click the Toolbars tab.

3. Click the built-in toolbar that you want to reset.

4. Click Reset.

5. Click OK.

6. Click Close.

A built-in toolbar was modified.

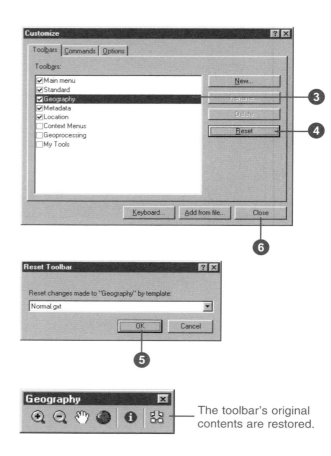

The toolbar's original contents are restored.

Changing a command's appearance

ArcCatalog lets you modify the appearance of buttons, tools, and menus without programming. You can add a grouping bar to visually separate commands used for different tasks such as browsing and querying. You can also modify the command's display type—either "Image Only", "Image and Text", or "Text Only". By default, a button or tool dropped onto a toolbar has the display type "Image Only"; when dropped onto a menu their display type is "Image and Text". Menus can only have the display type "Text Only".

Changing a command's caption changes the text that appears with the appropriate display types. Menus and their contents can be accessed from the keyboard by holding down the Alt key and pressing the underlined letter. To create an access key, type an ampersand (&) in front of a letter in the command's caption.

Other properties, such as a button's ToolTip and Message, can only be modified with programming. When you hold the mouse pointer over a ►

Grouping commands

1. Show the toolbar containing the commands that you want to group together.

2. Click the Tools menu and click Customize.

3. In the toolbar, right-click the command located to the right of where the grouping bar should be placed.

4. Check Begin a Group to show a grouping bar to the left of a command.

   Uncheck Begin a Group to remove a grouping bar.

5. Click Close in the Customize dialog box.

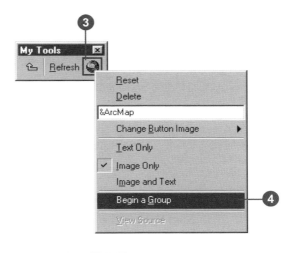

A grouping bar appears in the toolbar to the left of the command.

Changing the display type

1. Show the toolbar containing the command whose display type you want to change.

2. Click the Tools menu and click Customize.

3. In the toolbar, right-click the command you want to change. ►

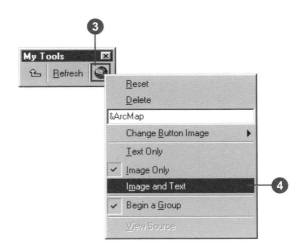

command, its *ToolTip*—a short message in a floating yellow box—displays. A command's Message displays in the status bar. The Message describes the action performed by the command.

4. Check Text Only to display only the command's caption.

   Check Image Only to display only the command's image.

   Check Image and Text to display both its image and its caption.

   The command's display type changes.

5. Click Close in the Customize dialog box.

The command's display type changes.

Changing the image

1. Show the toolbar containing the command whose image you want to change.

2. Click the Tools menu and click Customize.

3. In the toolbar, right-click the command you want to change.

4. Point to Change Button Image.

5. Click one of the images displayed. Or click Browse, navigate to a custom image, then click Open.

   The new image is applied. It appears in the toolbar if the display type is Image Only or Image and Text.

6. Click Close in the Customize dialog box.

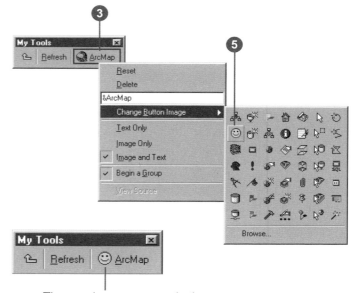

The new image appears in the toolbar if the display type is Image Only or Image and Text.

Changing the caption

1. Show the toolbar containing the command whose caption you want to change.

2. Click the Tools menu and click Customize.

3. In the toolbar, right-click the command you want to change.

4. Type a new caption in the edit box on the context menu.

5. Press Enter.

   The new caption is applied. It appears in the toolbar if the display type is Text Only or Image and Text.

6. Click Close in the Customize dialog box.

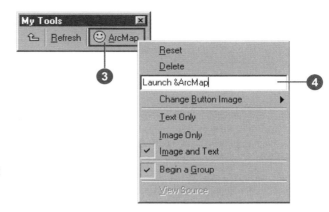

The new caption appears in the toolbar if the display type is Text Only or Image and Text.

Resetting a built-in command

1. Show the toolbar with the command you want to reset.

2. Click the Tools menu and click Customize.

3. In the toolbar, right-click the command.

4. Click Reset.

   The command's image, caption, and display type return to their default settings.

5. Click Close in the Customize dialog box.

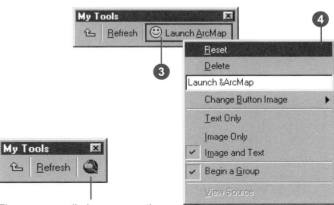

The command's image, caption, and display type return to their default settings.

Creating shortcut keys

When you access a menu from the keyboard using an access key, the menu opens and you can see its contents. In contrast, when you use a command's *shortcut key*, you can execute the command directly without having to open and navigate the menu first. For example, Ctrl + C is a well-known shortcut for copying something in Windows. One command can have many shortcuts assigned to it, but each shortcut can only be assigned to one command. A command's first shortcut is displayed to its right when the command appears in a menu.

Assigning a shortcut key

1. Click the Tools menu and click Customize.

2. Click Keyboard.

3. Click the category containing the command you want to modify.

4. Click the command to which you want to add a shortcut key.

5. Click in the Press new shortcut key text box, then press the keys on the keyboard that you want to use for a shortcut.

   If those keys have been assigned to another command, that command's name will appear below.

6. Click Assign if the keys aren't currently assigned to another command.

   The new shortcut appears in the Current Key/s list.

7. Click Close in the Customize Keyboard dialog box.

8. Click Close in the Customize dialog box.

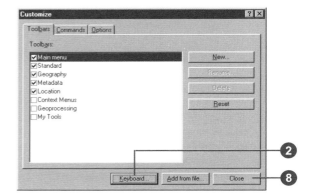

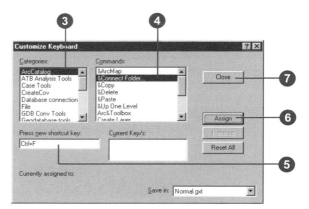

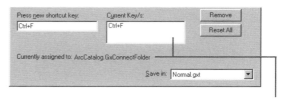

The new shortcut key appears in the Current Key/s list and is assigned to the command.

Removing a shortcut key

1. Click the Tools menu and click Customize.

2. Click Keyboard.

3. Click the category that contains the command you want to modify.

4. Click the command from which you want to remove a keyboard shortcut.

5. Click the shortcut in the Current Key/s list that you want to delete.

6. Click Remove.

7. Click Close in the Customize Keyboard dialog box.

8. Click Close in the Customize dialog box.

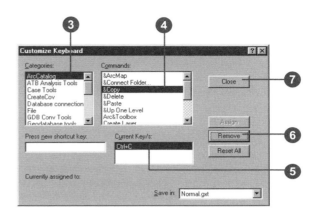

Resetting built-in shortcut keys

1. Click the Tools menu and click Customize.

2. Click Keyboard.

3. Click Reset All.

   Click Yes when asked if you want to reset your shortcuts.

4. Click Close in the Customize Keyboard dialog box.

5. Click Close in the Customize dialog box.

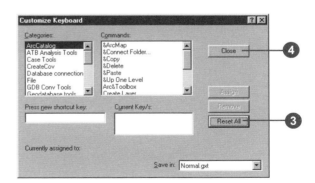

Creating and running macros

ArcCatalog comes with Visual Basic for Applications (VBA). VBA is not a standalone program. It provides an integrated programming environment, the Visual Basic Editor (VBE), that lets you write a VB macro and then debug and test it right away in the Catalog. A macro is a few lines of code that run in the Catalog; for example, the macro might analyze the currently selected coverage. A macro can integrate some or all of VB's functionality, such as using message boxes for input, with the functionality available in ArcCatalog.

When you create a macro, you're creating a VB Sub procedure. The procedure's name is the name you assign to the macro. You add code to the procedure in a Code window just as you would in VB. When you create a new macro in the Macros dialog box, precede the macro's name with the name of the module to store it in. To save your macro as part of the Catalog, you would type a name such as "ThisDocument.myMacro". You can also organize your macros into different modules; each module has its own Code ▶

Creating a macro

1. Click the Tools menu, point to Macros, then click Macros.

2. Type the name of the macro you want to create in the Macro name text box.

   To place the macro in a specific location, type ThisDocument or the module's name and then a period (.) before the new macro's name. Otherwise, the macro will be added to the "NewMacros" module.

3. Click Create or press Enter.

   The Code window appears containing a stub for the new Sub procedure.

4. Type the code for the macro.

5. Click the VBE File menu and click Save Project.

6. Click the Close button in the VBE.

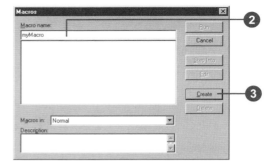

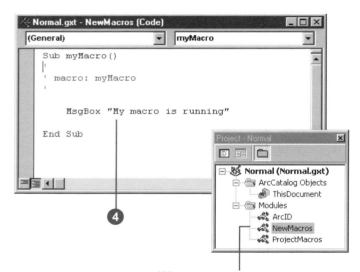

When you type a macro name without specifying which module to put it in, the macro is placed in the NewMacros module.

window. To add your macro to a specific module, type the module name before the macro's name, for example, "ProjectMacros.getSelected". If the module doesn't already exist, a new module with that name is created for you and added to the VBE project. Similarly, if you provide a name for a new macro but don't specify which module to store it in, a new module is created called "NewMacros". Using modules makes it easier to share your VB code with others; you can export a module to a .bas file from, and import a .bas file to, your VBE project. For more information about creating macros, see *Exploring ArcObjects* or the ArcObjects™ Developer Help system.

Editing a macro

1. Click the Tools menu, point to Macros, then click Macros.

2. In the list below the Macro name text box, click the name of the macro you want to edit.

3. Click Edit.

   The code that's been written for the macro appears in the Code window.

4. Edit the code.

5. Click the VBE File menu and click Save Project.

6. Click the Close button in the VBE.

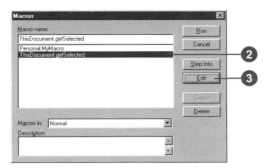

Adding a macro to a toolbar or menu

1. Show the toolbar to which you want to add a macro.

2. Click the Tools menu and click Customize.

3. Click the Commands tab.

4. Click Macros in the Categories list.

5. Click and drag the macro from the Commands list and drop it on the toolbar.

   The macro appears in the toolbar or menu.

6. Click Close.

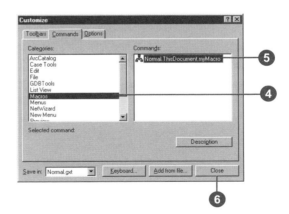

The macro appears in the toolbar.

Running a macro in the Macros dialog box

1. Click the Tools menu, point to Macros, and click Macros.

2. Click the Macro that you want to run.

3. Click Run.

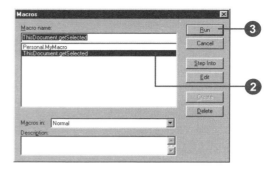

Running a macro in the Visual Basic Editor

1. Click the Tools menu, point to Macros, and click Visual Basic Editor.

2. In the VBE Project window, double-click ThisDocument or the module containing the macro that you want to run.

   The Code window for that module appears.

3. Position the cursor inside the appropriate Sub procedure.

4. Click the VBE Run menu and click Run Sub/UserForm.

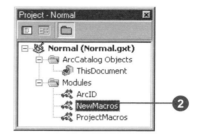

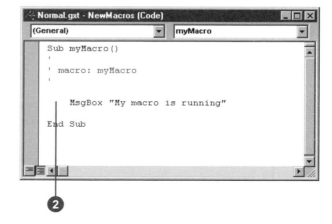

Creating custom commands with VBA

ArcCatalog uses Automation, which is a feature of the Component Object Model (COM) technology; it lets you access the Catalog's objects in VB and other languages, tools, and applications that support Automation. For example, you can analyze a data source with VB using ArcCatalog objects but without using ArcCatalog itself.

Toolbars and commands are COM objects, too. You can create custom objects in VBA. To be a command, the object must meet a basic set of requirements for all commands. To be a button, it must also satisfy the button requirements. *Exploring ArcObjects*, the ArcObjects Developer Help topics, and the VBE online Help describe the interfaces, methods, events, and properties that are available in VBA. The ArcCatalog customization environment makes it easy to create custom commands with VBA. You can create a new button, tool, combo box, or edit box (collectively called UIControls) in the Customize dialog box, then attach behavior that incorporates the Catalog's objects.

1. Show the toolbar to which you want to add a new command.

2. Click the Tools menu and click Customize.

3. Click the Commands tab.

4. Click UIControls in the Categories list.

5. Click New UIControl.

6. Click the type of UIControl you want to create.

7. Click Create to create the control without attaching code to it. The name of the control appears in the Commands list. You can add code for the control at another time.

   If you want to start adding code to the control right away, click Create and Edit and skip to step 11. ►

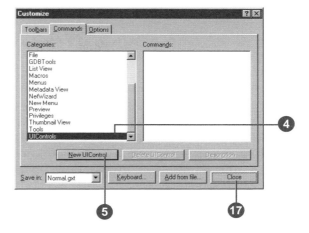

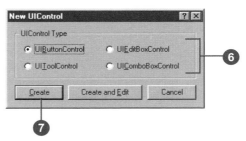

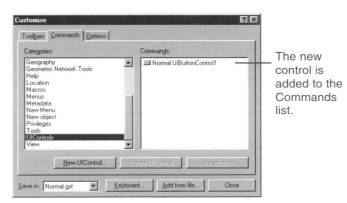

The new control is added to the Commands list.

8. Click and drag the newly created UIControl and drop it on a toolbar or menu.

9. In the toolbar, right-click the control and set its image, caption, and other properties.

10. Right-click the new control and click View Source.

   The Visual Basic Editor appears, displaying the control's code in the Code window.

11. Click the dropdown arrow and click one of the control's event procedures.

12. Type code for the event procedure.

13. Repeat steps 11 and 12 until all the appropriate event procedures have been coded.

14. Click the VBE File menu and click Save Project.

15. Click the Close button in the Visual Basic Editor.

16. If you clicked Create and Edit in step 7, open the Customize dialog box, click the Commands tab, and drag the newly created UIControl from the commands list to a toolbar or menu.

17. Click Close in the Customize dialog box.

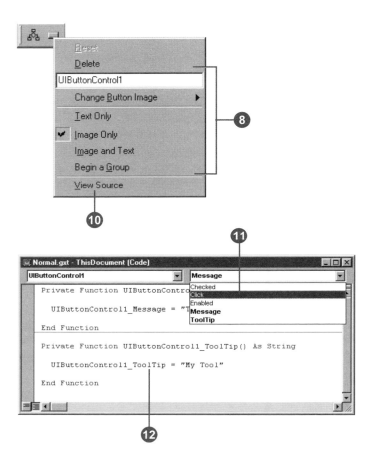

Working with UIControls

If you create a macro and add it to a toolbar, you've essentially customized what happens when you click the button. UIControls, however, let you create custom commands with VBA that work similarly to the buttons, tools, combo boxes, and edit boxes that come with ArcCatalog.

Typically in ArcCatalog, you use buttons to start, end, or interrupt an action or a series of actions. When you create a button, you write code that sets its properties including its ToolTip, the message that appears in the status bar describing what the button does, and whether its enabled. You also define the code that responds to the button's click event; this is the code that runs when you click the button on a toolbar.

Though similar to buttons, tools let you interact with the display—for example, you can zoom in or pan around your map in Geography view. In addition to setting the tool's properties, your code can respond to mouse and key events (clicking a mouse button or pressing a key) and to someone selecting, double-clicking, or right-clicking the tool. A tool can respond when the display refreshes or when it is deactivated.

A combo box combines the features of a text box and a list box. You can use combo boxes to provide a set of items to choose from, or you can let someone type a value that isn't in the list. You can add or remove entries from the list using the combo box. Its properties let you work with the selected item or the text in the edit box, as well as determine how many entries are in the list. The combo box can respond to several events such as changing the current selection or changing the text in the edit box. As with

buttons and tools, you can also set the control's ToolTip and provide a status bar message.

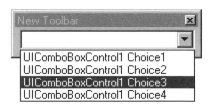

Edit boxes let you display information entered by the user or data derived from an external source. The Clear method removes the contents of the edit box, while the Text property provides access to the text that's displayed. You can specify whether or not the control is enabled and have the tool respond when someone changes the text or presses a key. You can set the control's ToolTip and its status bar message.

Writing code for a command's event procedures lets the command respond to user interaction or the current state of the Catalog. And, by using combo and edit boxes in ArcCatalog, you might be able to avoid using a dialog box or UserForm to get information. UIControls let you create sophisticated custom commands that make ArcCatalog an efficient and powerful way to create and finish your projects.

Adding custom commands

You don't have to use VBA to create custom commands. In fact, in some cases, your custom commands may require you to use another development environment. You can create custom objects in any programming language that supports COM; see *Exploring ArcObjects* for details. Custom commands or toolbars created outside VBA are often distributed as ActiveX libraries (DLLs). Before you can add a custom command to the Catalog, you must ensure that either you or the installation process by which you acquired it registers its ActiveX DLL. After registering the file on your computer, you must register the custom objects it contains with ArcCatalog. You can add the custom command to any toolbar or menu.

1. Click the Tools menu and click Customize.

2. Click Add from file.

3. Navigate to the file containing the custom command.

4. Click the file and click Open.

   The Added Objects dialog box appears, reporting which new objects have been registered with ArcCatalog.

5. Click OK.

   The custom commands appear in the Commands list for the appropriate category; a new category may be added to the Categories list.

6. Click Close in the Customize dialog box.

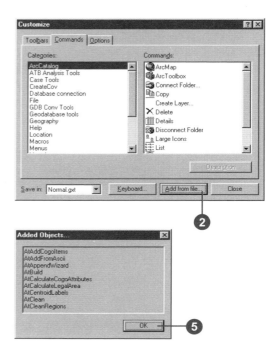

The custom commands appear in the Commands list for the appropriate category; a new category may be added to the list.

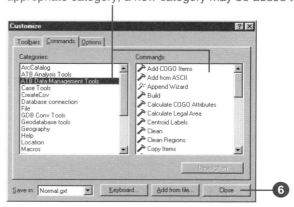

Updating the ArcID module

If you write macros that use COM objects, the ArcID module will be useful. You can refer to COM objects by name when using a method such as CommandBars.Find. If you do this, a list of the commands documented in the ArcID module will appear with Visual Basic's code completion feature. After adding objects to the Catalog from a file, update the ArcID module so that the newly added commands appear in the list along with the commands that are built into the Catalog. After updating the module, commands that have been removed will no longer appear in the list.

1. Click the Tools menu and click Customize.

2. Click the Options tab.

3. Click Update ArcID module.

4. Click Close.

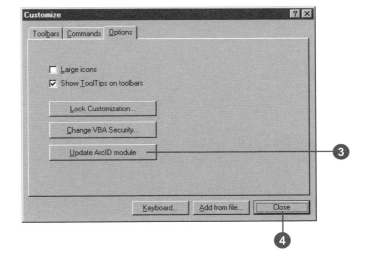

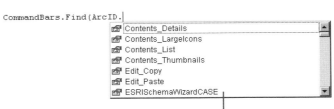

Using VB's code completion feature, you can list the commands that have been added to the ArcID module.

Changing VBA security

A macro virus is a type of computer virus that's stored in a macro or an add-in. When you open the file or perform an action that triggers a macro virus, the macro virus might be activated, transmitted to your computer, and stored as part of the Catalog. From that point on, every document you open or every file you save could be automatically "infected" with the macro virus; if others open these infected documents, the macro virus is transmitted to their computers. ESRI applications offer the levels of security described in the Security dialog box to reduce the chances of macro viruses infecting your documents, files, and add-ins.

Tip

Locking customizations
Since ArcCatalog is designed to be a personal data browser and manager, it doesn't use templates the way ArcMap does. To learn how to lock your customizations in the Catalog's Normal template or in ArcMap, see Using ArcMap.

1. Click the Tools menu and click Customize.

2. Click the Options tab.

3. Click Change VBA Security.

4. Click the level of security you want.

5. Click the Trusted Sources tab to see a list of the names of organizations or individuals whose signed macros will be allowed to run.

6. Click OK.

7. Click Close.

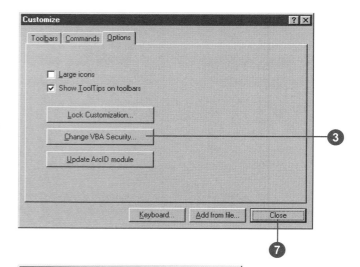

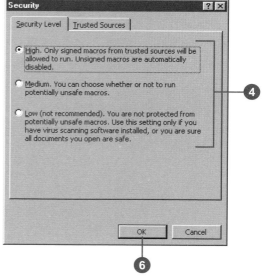

Using the ArcObjects Developer Help system

When you start creating macros and UIControls and programming how they work, you'll find lots of useful information in the ArcObjects Developer Help system. In addition to a general overview about how to get started, it provides object model diagrams, sample code, and detailed technical documents that will help you. For more indepth information about how to customize ArcCatalog, read *Exploring ArcObjects*.

The ArcObjects Class Help section of the Developer Help system also contains details about how to use each interface, property, and method that is available with ArcObjects. You can access this information where you need it most—while you are writing code. For example, you can select the method you are interested in and press F1.

1. Click the Start button on the Windows taskbar.

2. Point to Programs.

3. Point to ArcGIS.

4. Click ArcObjects Developer Help.

   The ArcObjects Developer Help system appears.

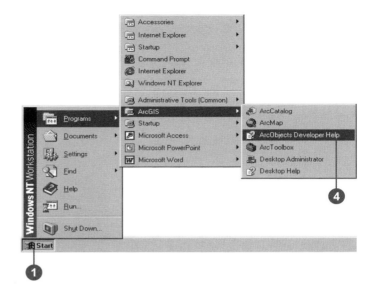

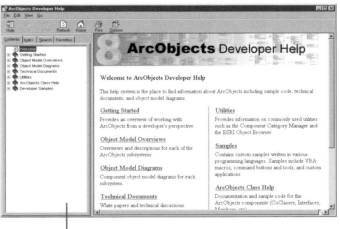

The ArcObjects Developer Help system appears.

Glossary

access key

An access key lets you access the contents of the Main Menu from the keyboard by holding down the Alt key and then pressing the underlined letter on the menu or menu item. You create an access key by placing an ampersand (&) in front of the appropriate letter in the command's caption.

active data frame

The data frame in a map that is currently being worked on; for example, the data frame to which layers are being added. The active data frame is shown in bold text in the table of contents.

AML

ARC Macro Language. A high-level algorithmic language for generating end-user applications. Features include the ability to create onscreen menus, use and assign variables, control statement execution, and get and use map or page unit coordinates. AML includes an extensive set of commands that can be used interactively or in AML programs (macros), as well as commands that report on the status of ArcInfo environment settings. ArcToolbox can be used to execute AML programs ("Run Geoprocessing AML" tool) and AML scripts for later use.

annotation

1. Descriptive text used to label coverage features. It is used for display, not for analysis.

2. A feature class used to label other features. Information stored for annotation includes a text string, the location at which it is displayed, and a text symbol (color, font, size, etc.) for display.

ArcInfo LIBRARIAN

A set of software tools to manage and access large geographic datasets in a map library. ArcInfo LIBRARIAN commands create and define a map library, move data in and out of a library, query the data in a map library, and display the results of a query.

ArcStorm database

An ArcStorm database is a collection of libraries, layers, INFO tables, and external database management system (DBMS) tables. Data stored in an ArcStorm database benefits from the transaction management and data archiving capabilities of ArcStorm.

attribute

1. A characteristic of a geographic feature described by numbers, characters, dates, and images and typically stored in tabular format. For example, the attributes of a well might include depth and gallons per minute.

2. A column in a dataset's table. See also column, field, and item.

attribute table

A table containing rows and columns. Attribute tables are associated with a class of geographic features such as wells or roads. Each row represents a geographic feature. Each column represents one attribute of a feature, with the same column representing the same attribute in each row. See also feature attribute table.

Automation

Automation is a feature of the COM technology. It lets you access ArcObjects in VB and in other languages, tools, and applications that support Automation. For example, with VB you can access the data in and properties of the selected item in the Catalog tree.

batch mode operation

Executes a given ArcToolbox tool two or more times on the items listed in the batch table. You can save the contents of the batch table as a geoprocessing AML and reload the AML later for execution.

batch table

Displays the input name, user-selected parameters, and the output name, where applicable, for all items to be processed by the ArcToolbox tool.

button

A command that runs a macro or custom code when clicked. Buttons can be added to any menu or toolbar. When they appear in a menu, buttons are referred to as menu items.

CAD

Computer-aided design. An automated system for the design, drafting, and display of graphically oriented information.

CAD dataset

A feature dataset representation of a CAD drawing. A CAD feature dataset comprises feature classes representing all the points, lines, polygons, or annotation in the CAD drawing. For example, a CAD drawing may contain two line layers representing roads and parcel boundaries, respectively. The CAD dataset's line feature class represents all features in both the road and parcel boundary layers.

CAD drawing

1. The digital equivalent of a drawing, figure, or schematic created using a CAD system. For example, a drawing file or DWG file in AutoCAD®.

2. An item in the Catalog tree representing all features and annotation in a CAD file. When you add a CAD drawing to a map or preview its contents in the Catalog, features are drawn using the symbology defined in the CAD file. You can't analyze the features with this representation; you must use the appropriate feature class in the CAD dataset.

caption

The text for a command that appears with the "Text Only" and "Image and Text" display types. For example, the name of a menu or menu item.

cardinality

A property of a relationship. Cardinality describes how many objects of type A are associated with how many objects of type B. Relationships can have many-to-one, one-to-one, one-to-many, and many-to-many cardinalities.

Catalog tree

Contains a set of folder connections that provide access to geographic data stored in folders on local disks or shared on the network. It also includes folders that let you manage database connections and coordinate systems. The Catalog tree provides a hierarchical view of the geographic data in those folders.

cell

See raster cell.

color ramp

A range of colors used in a map to show ranking or order of feature attributes.

column

The vertical dimension of a table. A column has a name and a data type applied to all values in the column. See also item, field, and attribute.

COM

Component Object Model. The Component Object Model is a technology, not a programming language. COM is a binary specification that establishes a common way of building software components. COM objects have interfaces that contain methods and properties.

combo box

A command that combines the features of an edit box and a list box. A combo box lets you type information or choose an option from a list. For example, the Location combo box in ArcCatalog lets you select an item in the Catalog tree by typing its path or choosing a path from its dropdown list.

command

An instruction that, when issued by the user, causes ArcMap or ArcCatalog to perform an action. A command can be typed from a keyboard, selected from a menu or toolbar, or embedded in program code.

composite relationship

Composite relationships describe associations where the lifetime of one object controls the lifetime of its related objects. An example is the association between highways and points for placing a highway shield marker. Shield points can't exist without a highway. See also relationship, simple relationship.

Content Standard for Digital Geospatial Metadata

A metadata style guide authored by the FGDC. All U.S. government agencies and state and local agencies that receive federal funds to create metadata must create metadata following this standard.

coordinate

A set of numbers that designate location in a given reference system such as x,y in a planar coordinate system or x,y,z in a three-dimensional coordinate system. Coordinates represent locations on the earth's surface relative to other locations. See also vector.

coordinate system

1. A reference system used to measure horizontal and vertical distances on a planimetric map. A coordinate system is usually defined by a map projection, a spheroid of reference, a datum, one or more standard parallels, a central meridian, and possible shifts

in the x- and y-directions to locate x,y positions of point, line, and area features.

2. An item in ArcCatalog representing a projection file, which contains the parameters defining a coordinate system. The contents of the projection file can either be in the format defined by ArcInfo Workstation or in the format defined by ArcCatalog.

coverage

A vector data storage format for storing the location, shape, and attributes of geographic features. A coverage usually represents a single theme, such as soils, streams, roads, or land use. One of the primary vector data storage formats for ArcInfo.

A coverage stores geographic features as primary features (such as arcs, nodes, polygons, and label points) and secondary features (such as tics, map extent, links, and annotation). Associated feature attribute tables describe and store attributes of the geographic features.

coverage units

The units (for example, feet, meters, inches) of the coordinate system in which a coverage is stored.

dangle tolerance

The minimum length allowed for dangling arcs during the CLEAN process. CLEAN removes dangling arcs that are shorter than the dangle length. Also known as the dangle length.

dangling arc

An arc having the same polygon on both its left and right sides and having at least one node that does not connect to any other arc. It often identifies where a polygon does not close properly (for example, undershoot), where arcs don't connect properly, or where an arc was digitized past its intersection with another arc (i.e., overshoot). A dangling arc is not always an error. For example, dangling arcs can represent cul-de-sacs in street centerline maps.

data

A collection of related facts usually arranged in a particular format and gathered for a particular purpose.

data frame

A frame on the map that displays layers occupying the same geographic area. You may have one or more data frames on your map depending on how you want to organize your data. For instance, one data frame might highlight a study area, and another might provide an overview of where the study area is located.

data source

Any geographic data, such as a coverage, shapefile, raster, or feature class, in a geodatabase.

data type

The characteristic of columns and variables that defines what types of data values they can store. Examples include character, floating point, and integer.

data view

An all-purpose view for exploring, displaying, and querying geographic data in ArcMap. This view hides all map elements such as titles, North arrows, and scale bars. See also layout view.

database

1. A collection of related files organized for efficient retrieval of information. A logical collection of interrelated information, managed and stored as a unit, usually on some form of mass-storage system such as magnetic tape or a disk. A GIS database includes data about the spatial location and shape of geographic features recorded as points, lines, areas, pixels, grid cells, or TINs, as well as their attributes.

2. In Sybase® and SQL Server RDBMSs, tables and other objects in the database are organized into different databases; for example, each database might contain data for different projects. When connecting to these RDBMSs, you must provide the name of the database containing the objects you want to access in addition to the name of the server or IP address and instance or port number.

database connection

A connection in ArcCatalog to a database. Database connections have a state—they are either connected to or disconnected from the database. If you delete a database connection, you only delete the connection itself, not the database or its contents. In general, when you create a database connection, you choose a data provider that will retrieve your data from the appropriate database. To work with the features in a multiuser geodatabase, use ArcSDE as the data provider. OLE DB providers let you access data in many different databases.

database management system

A set of computer programs for organizing the information in a database. A DBMS supports the structuring of the database in a standard format and provides tools for data input, verification, storage, retrieval, query, and manipulation.

dataset

1. A feature dataset in a geodatabase.

2. Any geographic data, such as a coverage, shapefile, raster, or a feature class, in a geodatabase. Also known as a data source.

datum

A set of parameters and control points used to accurately define the three-dimensional shape of the earth (for example, as a spheroid). The datum defines a geographic coordinate system that is the basis for a planar coordinate system. For example, the North American Datum for 1983 (NAD83) is the datum for map

projections and coordinates within the United States and throughout North America.

decimal degrees

Degrees of latitude and longitude expressed as a decimal rather than in degrees, minutes, and seconds.

DEM

Digital elevation model.

1. A digital representation of a continuous variable over a two-dimensional surface by a regular array of z-values referenced to a common datum. Digital elevation models are typically used to represent terrain relief. Also referred to as digital terrain model (DTM).

2. An elevation database for elevation data by map sheet from the National Mapping Division of the U.S. Geological Survey (USGS).

3. The format of the USGS digital elevation datasets.

destination

The secondary object in a relationship. For example, a table containing attributes that are associated with features in a feature class. See also relationship and origin.

digitize

1. To encode geographic features in digital form as x,y coordinates.

2. The process of using a digitizer to encode the locations of geographic features by converting their map positions to a series of x,y coordinates stored in computer files. Pushing a digitizer button records an x,y coordinate. A digitized line is created by recording a series of x,y coordinates.

digitizer

A device that consists of a table and a cursor with crosshairs and keys used to digitize geographic features.

directory

A computer term identifying a location on a disk containing a set of files and other directories (subdirectories). Operating systems use directories to organize data. The location of a directory is specified with a path.

disk

A storage medium consisting of a spinning disk coated with a magnetic material for recording digital information.

display type

A command's display type controls whether you see its image, its caption, or both when it appears on a toolbar or in a menu. Menus show text only; similarly, you can't change the display type for combo box and edit box commands.

documentation

Text in an item's metadata describing its contents, for example, where the data came from or a description of the values contained in an attribute. Documentation must be provided by a person, for example, using the Catalog's metadata editor.

double precision

Refers to a high level of coordinate accuracy based on the possible number of significant digits that can be stored for each coordinate. ArcInfo datasets can be stored in either single- or double-precision coordinates. Double-precision coverages store up to 15 significant digits per coordinate (typically 13 to 14 significant digits), retaining the accuracy of much less than one meter at a global extent. See also single precision.

edit box

A command that displays text typed by a person or derived from another source. For example, in ArcMap you can type the scale at which you want to see the map.

enclosures

Files describing the contents of a data source can be enclosed in the metadata. A copy of the file is contained within the metadata. Enclosing files in metadata works the same way that you enclose files in an e-mail message.

event

Objects execute code in response to some event. Events are caused by user interaction with an application. For example, clicking a button and closing a form trigger events. In structured programs, the program has control and asks the user for information needed to proceed. In event-driven programming, the user is in control of the program and causes events when a program's services are required.

extent

The coordinates defining the minimum bounding rectangle (i.e., xmin, ymin and xmax, ymax) of a data source. All coordinates for the data source fall within this boundary.

feature

1. An object class in a geodatabase that has a field of type geometry. Features are stored in feature classes.

2. A representation of a real-world object.

feature attribute table

A table used to store attribute information for a specific coverage feature class. ArcInfo maintains the first several items of these tables. Feature attribute tables supported for coverages include the following:

<cover>.PAT   for polygons or points
<cover>.AAT   for arcs
<cover>.NAT   for nodes
<cover>.RAT   for routes
<cover>.SEC   for sections
<cover>.PAT   for regions
<cover>.TAT   for annotation (text)

where <cover> is the coverage name.

feature class

1. A classification describing the format of geographic features and supporting data in a coverage. Coverage feature classes for representing geographic features include point, arc, node, route-system, route, section, polygon, and region. One or more coverage features are used to model geographic features; for example, arcs and nodes can be used to model linear features such as street centerlines. The tic, annotation, link, and boundary feature classes provide supporting data for coverage data management and viewing.

2. The conceptual representation of a geographic feature. When referring to geographic features, feature classes include point, line, area, and annotation. In a geodatabase, an object class that stores features and has a field of type geometry.

3. The collection of all the point, line, or polygon features or annotation in a CAD dataset.

feature dataset

A collection of feature classes that share the same spatial reference. Because the feature classes share the same spatial reference, they can participate in topological relationships with each other such as in a geometric network. Object classes and relationship classes can also be stored in a feature dataset.

FGDC

Federal Geographic Data Committee. An interagency committee, organized in 1990, that promotes the coordinated use, sharing, and dissemination of geospatial data on a national basis. The FGDC is composed of representatives from 16 Cabinet-level and independent U.S. federal agencies. The FGDC authored the *Content Standard for Digital Geospatial Metadata.*

field

A column in a table. Each field contains the values for a single attribute. See also item, attribute, and column.

file

A set of related information that a computer can access by a unique name (for example, a text file, a datafile, a DLG file). Files are the logical units managed on disk by the computer's operating system. Files may be stored on tapes or disks.

file type

Files that are not geographic data sources can appear in ArcCatalog if they have been added to the file types list. A file type consists of a description of the file's format such as "Text Document", the file extension associated with this type of file such as ".txt", and the icon used to represent these files.

folder

A location on a disk containing a set of files and other folders. See also directory.

folder connection

A folder connection provides access to the contents of an entire disk on your computer, a specific folder on a disk, or a shared folder on the network. When you remove a folder connection from ArcCatalog, the folder and its contents are not deleted, but you can no longer access its contents in ArcCatalog.

foreign key

1. One or more table attributes can uniquely identify related records in another table. A foreign key is the primary key of

another table. Foreign key–Primary key relationships define a relational join.

2. Used to create a relationship class, the foreign key resides in the destination object class. To join two object classes together, the primary key and foreign key must share the same values.

See also primary key, relate, and relationship class.

format

The pattern into which data is systematically arranged for use on a computer. A file format is the specific design of how information is organized in the file. For example, ArcInfo has specific, proprietary formats used to store coverages. DLG, DEM, and TIGER® are geographic datasets with different file formats.

freeze

When exploring a table's contents, you can fix a column in place at the left side of the table. When you scroll horizontally through the table's columns, the frozen column stays in place while the other columns scroll normally. For example, freeze the "Country_name" column so it's easy to see which country has which population or birth rate.

fuzzy tolerance

The fuzzy tolerance is an extremely small distance used to resolve inexact intersection locations due to the limited arithmetic precision of computers. It defines the resolution of a coverage resulting from the Clean operation or a topological overlay operation such as Union, Intersect, or Clip.

geodatabase

A geographic database that is hosted inside a relational database management system that provides services for managing geographic data. These services include validation rules, relationships, and topological associations.

geographic coordinates

A measurement of a location on the earth's surface expressed in degrees of latitude and longitude. See projected coordinates.

geographic data

The locations and descriptions of geographic features. The composite of spatial data and descriptive data.

geometric network

A geometric network can be thought of as a one-dimensional nonplanar graph, or logical network, that is composed of features. These features are constrained to exist within the network and can therefore be considered network features. ArcInfo 8 will automatically maintain the explicit topological relationships between network features in a geometric network.

Represent one-dimensional linear networks, such as a road system, a utility network, or a hydrologic network. Geometric networks contain feature classes that play a topological role in the network. These feature classes are homogeneous collections of one of these four network feature types: simple junction feature, complex junction feature, simple edge feature, and complex edge feature. More than one feature class can have the same type of network feature.

geoprocessing

GIS operations, such as geographic feature overlay, coverage selection and analysis, topology processing, and data conversion.

Geospatial Data Clearinghouse

Sponsored by the FGDC. A decentralized system of servers on the Internet that contain metadata describing available geographic data. This metadata follows the format specified by the *Content Standard for Digital Geospatial Metadata* to facilitate querying and enforce a consistent presentation across all servers.

GIS

Geographic information system. An organized collection of computer hardware, software, geographic data, and personnel designed to efficiently capture, store, update, manipulate, analyze, and display all forms of geographically referenced information.

grid

A geographic representation of the world as an array of equally sized square cells arranged in rows and columns. Each grid cell is referenced by its geographic x,y location. See raster.

group layer

Several layers that appear and act like an individual layer in the table of contents in ArcMap.

HTML

Hypertext markup language. An HTML file contains text and tags instructing an Internet browser application on how to present the text. For example, <B>24</B> will display the text "24" in bold.

HTML is well formed if all tags are properly closed and nested. For example, "<B>" is an opening tag, while "</B>" is a closing tag that must follow the text, which is to appear in bold. If the text should also be underlined, the appropriate "<UL>" tags must both appear either inside or outside of the "<B>" tags. Inline elements, which typically don't require closing tags, must also be properly closed; for example, "<HR/>" adds a line. Many applications that help you build Web pages do not generate well-formed HTML.

identify

To click a feature so that you can see its associated attributes.

image

A graphic representation or description of a scene, typically produced by an optical or electronic device. Common examples include remotely sensed data (for example, satellite data), scanned data, and photographs. An image is stored as a raster dataset of binary or integer values that represent the intensity of reflected light, heat, or other range of values on the electromagnetic spectrum. See also raster.

index

A special data structure used in a database to speed searching for records in tables or spatial features in geographic datasets. ArcInfo supports both spatial and attribute indexes.

INFO database

The contents of a set of INFO datafiles, feature attribute tables, and related files stored in each ArcInfo workspace under a subdirectory named INFO. This subdirectory contains all feature attribute tables for the set of coverages contained in the workspace. The INFO subdirectory doesn't appear in ArcCatalog.

IP address

A server's address on the network. The address consists of four numbers, each separated by a '.' (period).

ISO

International Standards Organization. A worldwide federation of national standards bodies (for example, ANSI from the United States). Among many other computing standards, ISO maintains an SQL standard.

item

1. A column of information in an attribute table, for example, a single attribute of a record in an INFO table.

2. An element in the Catalog tree. The Catalog tree can contain both geographic data sources and nongeographic elements, such as folders, folder connections, and file types.

join

The process of attaching tabular data to geographic features. Attributes in the table are appended to the features using key attributes. See also relational join.

key attributes

To join two objects together, each object must have a column containing the same values. For example, a country feature with a "Name" attribute can be joined to the appropriate record in a table of demographic data that has a "Country_Name" column. See also relate, relationship class, primary key, and foreign key.

label

Text added to a map to help identify features.

latitude–longitude

A spherical reference system used to measure locations on the earth's surface. Latitude and longitude are angles measured from the earth's center to locations on the earth's surface. Latitude measures angles in a north–south direction. Longitude measures angles in an east–west direction.

layer

A collection of similar geographic features—such as rivers, lakes, counties, or cities—of a particular area or place for display on a map. A layer references geographic data stored in a data source, such as a coverage, and defines how to display it, for example, draw major streams with a thick blue line and its tributaries with a thin blue line. You can create and manage layers as you would any other type of data in your database.

layout

The design or arrangement of elements—such as geographic data, North arrows, and scale bars—in a digital map display or printed map.

layout view

The view for laying out your map. Layout view shows the virtual page on which you place and arrange geographic data and map elements—such as titles, legends, and scale bars—for printing. See also data view.

library

A collection of spatially related ArcStorm or ArcInfo LIBRARIAN layers. A library has a spatial extent that applies to all layers in the library.

macro

A sequence of commands that can be executed as one command. Macros can be built to perform frequently used, as well as complex, operations. AML is used to create macros for ArcInfo. In ArcCatalog and ArcMap you use VBA to create macros.

map

1. A graphical presentation of geographic information. It contains geographic data and other elements, such as a title, North arrow, legend, and scale bar. You can interactively display and query the geographic data on a map and also prepare a printable map by arranging the map elements around the data in a visually pleasing manner.

2. A map is a document that lets you display and work with geographic data. A map contains one or more layers of geographic data and various supporting map elements such as scale bars. Layers on a map are contained in data frames. A data frame has properties, such as scale, projection, and extent, and also graphic properties such as where it is located on your map's

page. Some maps have one data frame, while other more advanced maps may have several data frames.

map document

The disk-based representation of a map. Map documents can be printed or embedded into other documents. Map documents have an .mxd file extension.

map projection

See projection.

map template

A kind of map document that provides a quick way to create a new map. Templates can contain data, a custom interface, and a predefined layout that arranges map elements, such as North arrows, scale bars, and logos, on the virtual page. Map templates have an .mxt file extension.

menu

A command that arranges other commands into a list.

menu item

See button.

metadata

1. Data about data. For GIS data, metadata usually means data that is designed to help a prospective user find GIS data, determine whether it will serve a particular purpose, obtain the data, and use it.

2. Metadata consists of properties and documentation. Properties are derived from the data source, while documentation is entered by a person. By default, ArcCatalog automatically creates and updates metadata, which is stored as well-formed XML data in a file alongside the data or within a geodatabase. Metadata for a

folder can also consist of a well-formed HTML file describing its contents.

metadata element

One piece of data within a data source's metadata. In the XML file, each element has an opening and closing tag. Elements can appear many times in an XML file; these are called repeating elements. For example, in a shapefile's metadata, an "<attr>" element might represent one of its attributes. Therefore, the "<attr>" element will appear many times—once for each attribute.

metadata profile

An extension to the FGDC's *Content Standard for Digital Geospatial Metadata*. Additional elements defined by ESRI are recorded in the ESRI Metadata Profile.

minimum bounding rectangle

A rectangle, oriented to the x- and y-axes, that bounds a geographic feature or a geographic dataset. It is specified by two coordinates: xmin, ymin and xmax, ymax. For example, the extent defines a minimum bounding rectangle for a coverage.

multiuser geodatabase

A versioned ArcSDE geodatabase that can handle multiple readers and writers.

NSDI

National Spatial Data Infrastructure. Coordinated by the FGDC. The NSDI encompasses policies, standards, and procedures for organizations to cooperatively produce and share geographic data. The NSDI is being developed in cooperation with organizations from state, local, and tribal governments; the academic community; and the private sector.

object class

While spatial objects (features) are stored in feature classes in a geodatabase, nonspatial objects are stored in object classes. A table is an object class if it has a column with the data type OID (Object Identifier), where each row in the table is an object. In a geodatabase, nonspatial objects can have custom behavior.

OGC

Open GIS Consortium. OpenGIS is defined as transparent access to heterogeneous geodata and geoprocessing resources in a networked environment. The goal is to provide a comprehensive suite of open interface specifications that enable developers to write interoperating components that provide these capabilities.

OLE DB provider

Object Linking and Embedding Database (OLE DB) provider. Each provider communicates with and retrieves data from a different database, but you can work with data retrieved by any provider the same way. Typically, they can only retrieve nonspatial data. However, if an OLE DB provider can retrieve geographic data in OpenGIS format, you can work with that data in ArcInfo.

origin

The primary object in a relationship. For example, a feature class containing points at which measurements are taken; the measurements are stored in the associated table. See also relationship and destination.

overshoot

That portion of an arc digitized past its intersection with another arc. See also dangling arc.

pan

To move the viewing window up, down, or sideways to display areas in a geographic dataset that, at the current viewing scale, lie outside the viewing window. See also zoom.

password

The password to use for authentication when you log in to the geodatabase.

path

The location of a file or directory on a disk. A path is always specific to the computer operating system.

path label

Describes the nature of the association between the objects in a relationship. The forward path label describes the relationship when navigated from the origin to the destination; for example, station points "have" measurements. The backward path label describes the same relationship negotiated from the destination to the origin, which might be "are taken at" in this example; measurements "are taken at" stations.

PC ARC/INFO coverage

A coverage created with PC ARC/INFO.

personal geodatabase

An ArcSDE database whose underlying DBMS is Microsoft Access.

pixel

See raster cell.

planar coordinate system

A two-dimensional measurement system that locates features on a map based on their distance from an origin (0,0) along two axes:

a horizontal x-axis representing east–west and a vertical y-axis representing north–south.

port number

The TCP/IP port number that the instance is communicating on.

pre-7.0 coverage

Coverages created with a version of ArcInfo prior to version 7 can't be accessed by ArcCatalog and ArcMap until after the ConvertWorkspace command has been used to modify the coverage workspace.

precision

Refers to the number of significant digits used to store numbers and, in particular, coordinate values. Precision is important for accurate feature representation, analysis, and mapping. ArcInfo supports single and double precision.

preliminary topology

Refers to incomplete region topology. Region topology defines region–arc and region–polygon relationships. A topological region has both the region–arc relationship and the region–polygon relationship. A preliminary region has the region–arc relationship but not the region–polygon relationship. In other words, preliminary regions have no polygon topology. Coverages with preliminary topology have red in their icons in the Catalog.

primary key

1. One or more attributes whose values uniquely identify a row in a database table.

2. Used to create a relationship class, the primary key resides in the origin object class. To join two object classes together, the primary key and foreign key must share the same values.

See also foreign key, relate, and relationship class.

projected coordinates

A measurement of locations on the earth's surface expressed in a two-dimensional system that locates features based on their distance from an origin (0,0) along two axes, a horizontal x-axis representing east–west and a vertical y-axis representing north–south. A map projection transforms latitude and longitude to x,y coordinates in a projected coordinate system. See also geographic coordinates.

projection

A mathematical formula that transforms feature locations from the earth's curved surface to a map's flat surface. A projected coordinate system employs a projection to transform locations expressed as latitude and longitude values to x,y coordinates. Projections cause distortions in one or more of these spatial properties: distance, area, shape, and direction.

projection file

A file that stores the parameters describing a coordinate system.

property

An attribute of an object defining one of its characteristics or an aspect of its behavior. For example, the Visible property affects whether a control can be seen at run time. You can set a data source's properties using its Properties dialog box.

pyramids

Reduced resolution layers, or pyramids, record the original data in a raster in decreasing levels of resolution. The coarsest level of resolution is used to quickly draw the entire dataset. As you zoom in, layers with finer resolutions are drawn; performance is maintained because you're drawing successively smaller areas.

query

A question or request used for selecting features. A query often appears in the form of a statement or logical expression. In ArcMap, a query contains a field, an operator, and a value.

raster

Represents any data source that uses the raster model to represent geographic information. See also grid and image.

raster band

A measure of some characteristic or quality of the features being observed in a raster. Some rasters have a single band; others have more than one. For example, satellite imagery commonly has multiple bands representing different wavelengths of energy from along the electromagnetic spectrum.

raster cell

A discretely uniform unit that represents a portion of the earth, such as a square meter or square mile. Each grid cell has a value that corresponds to the feature or characteristic at that site, such as a soil type, census tract, or vegetation class.

raster dataset

Contains raster data organized into bands. Each band consists of an array of cells with optional attributes for each cell (pixel).

raster model

A representation of the world as a surface divided into a regular grid of cells. Raster models are useful for storing data that varies continuously such as in an aerial photograph, a satellite image, a surface of chemical concentrations, or an elevation surface.

RDBMS

Relational database management system. A database management system with the ability to access data organized in tabular files

that can be related to each other by a common field. An RDBMS has the capability to recombine the data items from different files, providing powerful tools for data usage.

record

1. In an attribute table, a single "row" of thematic descriptors. In SQL terms, a record is analogous to a tuple.

2. A logical unit of data in a file. For example, there is one record in the ARC file for each arc in a coverage.

relate

An operation that establishes a temporary connection between corresponding records in two tables using an item common to both (i.e., key attributes). Each record in one table is connected to those records in the other table that share the same value for the common item. See also relationship class and relational join.

relational database

A method of structuring data as collections of tables that are logically associated to each other by shared attributes. Any data element can be found in a relation by knowing the name of the table, the attribute (column) name, and the value of the primary key. See also RDBMS, relate, key attributes, and relational join.

relational join

The operation of relating and physically merging two attribute tables using their common item.

relationship

An association or link between two objects in a database. Relationships can exist between spatial objects (features in feature classes), nonspatial objects (rows in a table), or between spatial and nonspatial objects.

relationship class

While spatial objects are stored in feature classes and nonspatial objects are stored in object classes, relationships are stored in relationship classes.

resolution

1. Resolution is the accuracy at which a given map scale can depict the location and shape of geographic features. The larger the map scale, the higher the possible resolution. As the map scale decreases, resolution diminishes and feature boundaries must be smoothed, simplified, or not shown at all. For example, small areas may have to be represented as points.

2. The size of the smallest feature that can be represented in a surface.

3. The number of cells in the x- and y-directions in a raster.

row

1. A record in an attribute table. The horizontal dimension of a table composed of a set of columns containing one data item each.

2. A horizontal group of cells in a raster.

scale

The relationship between the dimensions of features on a map and the geographic objects they represent on the earth, commonly expressed as a fraction or a ratio. A map scale of 1/100,000 or 1:100,000 means that one unit of measure on the map equals 100,000 of the same unit on the earth.

scale bar

A map element that shows the map scale graphically.

SDE

Spatial Database Engine™ (SDE). A high-performance spatial database manager that employs a true client/server architecture to perform efficient spatial operations and manage large, shared, geographic datasets.

server

The computer where the geodatabase you want to access is located.

service

In a geodatabase, the name of the process running on the ArcSDE server that allows connections and access to spatial data.

shapefile

A vector data storage format for storing the location, shape, and attributes of geographic features.

shortcut key

A command's shortcut key executes the command directly without first having to open and navigate a menu. For example, Ctrl+C is a well-known shortcut for copying a file in Windows.

simple relationship

Simple relationships describe associations between data sources that exist independently of each other. A coverage and table are independent of each other if when you delete the primary object, the related object continues to exist. For example, a table contains measurements taken at different stations. If you stop using a station and delete that point, you might keep the measurements for historical purposes.

single precision

Refers to a level of coordinate accuracy based on the number of significant digits that can be stored for each coordinate. Single-precision numbers store up to seven significant digits for each coordinate, retaining a precision of ±5 meters in an extent of 1,000,000 meters. ArcInfo datasets can be stored as either single- or double-precision coordinates. See also double precision.

single-user geodatabase

A personal geodatabase. It can handle a single writer and multiple readers.

snapping

The process of moving a feature to coincide exactly with coordinates of another feature within a specified snapping distance, or tolerance.

snapping tolerance

The distance within which the pointer or a feature will snap to another location. If the location being snapped to (vertex, boundary, midpoint, or connection) is within the distance you set, the pointer will automatically snap. For example, if you want to snap a power line to a utility pole and the snapping tolerance is set to 25 pixels, whenever the power line comes within a 25-pixel range of the pole, it will automatically snap to it. Snapping tolerance can be measured using either map units or pixels.

spatial data

The locations and shapes of geographic features with descriptions of each feature.

SQL

Structured Query Language. A syntax for defining and manipulating data from a relational database. Developed by IBM® in the 1970s, it has become an industry standard for query languages in most relational database management systems.

stylesheet

An XSL stylesheet, which selects data from an XML file, applies functions and formatting to the data and then specifies how to present the data.

subtypes

Although all objects in a feature class or object class must have the same behavior and attributes, not all objects have to share the same default values and validation rules. You can group features and objects into subtypes. Subtypes differentiate objects based on their rules.

surface

A geographic phenomenon represented as a set of continuous data such as elevation or air temperature over an area. A clear or sharp break in values of the phenomenon (breaklines) indicates a significant change in the structure of the phenomenon (for example, a cliff), not a change in geographic feature.

symbol

A graphic pattern used to represent a feature. For example, line symbols represent arc features; marker symbols, points; shade symbols, polygons; and text symbols, annotation. Many characteristics define symbols including color, size, angle, and pattern.

symbology

The criteria used to determine symbols for the features in a layer. A characteristic of a feature may influence the size, color, and shape of the symbol used.

table

A set of data elements that has a horizontal dimension (rows) and a vertical dimension (columns) in a database. A table has a specified number of columns but can have any number of rows.

table of contents

Lists all the data frames on the map and their layers.

tabular data

Descriptive information that is stored in rows and columns and can be linked to map features.

TCP/IP

Transmission Control Protocol (TCP) is a communication protocol layered above the Internet Protocol (IP). These are low-level communication protocols that allow computers to send and receive data.

text box

See edit box.

thumbnail

A snapshot describing the geographic data contained in a data source or layer, or a map layout. A thumbnail might provide an overview of all the features in a feature class or a detailed view of the features in, and the symbology of, a layer. Thumbnails are not updated automatically; they will go out of date if features are added to a data source or if the symbology of a layer changes.

tic

Registration or geographic control points for a coverage representing known locations on the earth's surface. Tics allow all coverage features to be recorded in a common coordinate system such as Universal Transverse Mercator (UTM). Tics are used to register map sheets when they are mounted on a digitizer

and to transform the coordinates of a coverage, for example, from digitizer units (inches) to the appropriate values for a coordinate system (which are measured in meters for UTM).

TIN

Triangulated irregular network. A surface representation derived from irregularly spaced sample points and breakline features. The TIN dataset includes topological relationships between points and their neighboring triangles. Each sample point has an x,y coordinate and a surface, or z-value. These points are connected by edges to form a set of nonoverlapping triangles used to represent the surface. TINs are also called irregular triangular mesh or irregular triangular surface model.

TIN dataset

A dataset containing a triangulated irregular network, a surface representation derived from irregularly spaced sample points and breaking features. The TIN dataset includes topological relationships between points and neighboring triangles.

tolerances

A coverage uses many processing tolerances (fuzzy, tic match, dangle length) and editing tolerances (weed, grain, edit distance, snap distance, and nodesnap distance). Stored in a TOL file, ArcInfo uses the values as defaults in many automation, editing, and processing operations. You can edit a coverage's tolerances using its Properties dialog box in ArcCatalog.

tool

1. An entity in ArcToolbox that performs a specific geoprocessing task such as generalizing lines. A tool can belong to two or more toolsets.

2. A command that requires interaction with the user interface before an action is performed. For example, with the Zoom In command, you must click or draw a box over the geographic data

or map before it is redrawn at a larger scale. Tools can be added to any toolbar.

toolbar

A set of commands that let you carry out related tasks. The Main Menu toolbar has a set of menu commands; other toolbars typically have a set of buttons. Toolbars can float on the desktop in their own window, or you can dock them at the top, bottom, or sides of the main window.

topology

The spatial relationships between connecting or adjacent coverage features (for example, arcs, nodes, polygons, and points). For example, the topology of an arc includes its from- and to-nodes and its left and right polygons. Topological relationships are built from simple elements into complex elements: points (simplest elements), arcs (sets of connected points), areas (sets of connected arcs), and routes (sets of sections, which are arcs or portions of arcs). Redundant data (coordinates) is eliminated because an arc may represent a linear feature, part of the boundary of an area feature, or both. Topology is useful in GIS because many spatial modeling operations don't require coordinates, only topological information. For example, to find an optimal path between two points requires a list of the arcs that connect to each other and the cost to traverse each arc in each direction. Coordinates are only needed for drawing the path after it is calculated.

tuple

A row in a relational table; synonymous with record, observation.

UIControl

A UIControl lets you create a custom command with VBA. There are four different types of UIControls: buttons, tools, combo boxes, and edit boxes.

undershoot

An arc that does not extend far enough to intersect another arc. See also dangling arc.

username

The identification to use for authentication when you log in to a geodatabase.

VBA

Visual Basic for Applications. The embedded programming environment for automating and customizing ESRI applications such as ArcMap and ArcCatalog. It offers the same powerful tools as Visual Basic (VB) in the context of an existing application and is the best option for customizing software that already meets most needs. By contrast, VB is a standalone tool for rapidly creating a special solution from scratch, be it an executable program, COM component, or an ActiveX control. An application that uses ArcMap or ArcCatalog may require the development of a COM component; consequently, in such instances, VB is the appropriate development environment.

vector

A coordinate-based data structure commonly used to represent linear geographic features. Each linear feature is represented as an ordered list of vertices. Traditional vector data structures include double-digitized polygons and arc–node models.

vector model

A representation of the world using points, lines, and areas. Vector models are useful for representing and storing discrete features such as buildings, pipes, or parcel boundaries.

version

A version is one alternative representation of a geodatabase that has an owner, a description, and a level of access (private, protected, and public).

views

Different ways to see the contents of the selected item in the Catalog tree.

workspace

A container of geographic data. This can be a folder that contains shapefiles, an ArcInfo workspace that contains coverages, a personal geodatabase, or a connection to a multiuser geodatabase.

XML

Extensible markup language. A markup language similar to HTML. With XML you define data using tags that add meaning. For example, <title>California geology</title> declares the text "California geology" to be a title, perhaps for a map. An XML file does not contain information about how to present the data. XML is well formed if an opening tag, such as "<title>", and a closing tag, such as "</title>", appear before and after each piece of data.

XSL

Extensible stylesheet language. XSL is a defined set of XML tags that can be used to query and evaluate XML data. XSL stylesheets are used in ArcCatalog to present metadata. See also stylesheet.

zoom

To enlarge and display greater detail of a portion of a geographic dataset.

Index

Folders (continued)
  shared on the network   9, 60, 76
Foreign key   40, 41, 202, 206, 265–266
Format
  changing a data source's   36, 62, 102, 126,
    127–128, 225–228
  defined   266
  supported   59, 60, 61, 62, 63, 64, 66, 68,
    69, 70, 71, 72, 73, 74, 86
Freezing columns
  defined   266
  in a table   17, 118
Fuzzy tolerance   187, 266

G

Geocoding   208, 215–216. *See also* geocoding
    services
Geocoding indexes   72, 208, 209
Geocoding services. *See also* items in
    ArcCatalog
  alternate street name table   212, 214, 215
  clientside   208, 209
  creating   208, 210–213
  defined   72, 208, 215–216
  input address fields   216–217, 220
  managing   84
  matching options
    intersection connectors   217, 222
    minimum candidate score   215–216,
      217, 221
    minimum match score   216, 217, 221
    spelling sensitivity   215, 217, 221
  output attributes   219, 224, 228
  output options   218, 223
  place name alias table   212–213, 214, 215
  potential candidates   215–216
  reference data   72, 208, 210, 211,
    214, 215

Geocoding services (continued)
  rematching addresses   229, 229–230,
    231–232, 232
  serverside   208, 209, 214
  styles   208, 210, 215
  using   225–228
Geocoding Services folder
  defined   72, 208
  hiding and showing   11, 84
Geodatabases. *See also* items in ArcCatalog
  connecting to   77–78, 78, 80, 81
  creating   100
  defined   12, 64, 266
  disconnecting from   80
  feature classes   64, 65, 105, 265
  feature datasets   64, 65, 265
  geocoding services   72, 208, 209
  geometric networks   64, 65, 266
  importing and exporting data   102
  multiuser   64, 65, 269
  owner   64, 66
  personal   12, 64, 65, 100, 270
  rasters   66
  relationship classes   64, 65
  single-user   12, 64, 65, 274
  subtypes   105, 274
  tables   64, 65
  version   78, 277
Geographic coordinate systems   74. *See also*
    latitude–longitude
Geographic coordinates   266
Geographic data   12, 266. *See also* data
    sources; Geography view
Geographic information system. *See* GIS
Geography view
  creating thumbnails   24, 110
  defined   14, 104
  identifying   16, 109, 154, 267
  panning   108, 270
  stop drawing data   107
  zooming   107, 108

Geometric networks   64, 65, 266. *See also*
    data sources
Geoprocessing   35, 266
Geospatial Data Clearinghouse   131, 266. *See
    also* NSDI
GIS   6, 267
Graphs   61, 106. *See also* items in ArcCatalog
Grids   66, 267. *See* raster
Group layers. *See also* items in
    ArcCatalog; layers
  adding and removing layers   163
  creating   163, 164
  defined   61, 160–161, 267
  previewing   16, 106

H

Help
  accessing topics   54
  ArcObjects   257
  contacting ESRI   6
  developer   257
  finding topics containing specific words   56
  in a dialog box   53, 139
  in the ArcCatalog window   53
  in the status bar   243–244, 253
  printing topics   55
  using the index   55
  What's This?   53, 139
HTML   18, 70, 134, 267
Hypertext markup language. *See* HTML

I

Image services   71. *See also* Internet services
Images   2, 66, 267. *See* raster
Indexes
  attribute   37, 174, 196–197, 201
  defined   267
  spatial   174, 175

Metadata. *See also* FGDC
    changing its appearance 19, 70, 131–132,
        134
    creating 3, 23, 110, 133, 134, 135, 136
    customizing 134, 137, 143
    defined 18, 70, 133, 269
    documentation 18, 43–44, 133, 137,
        140, 264
    editing 43–44, 133, 134, 137, 138–139,
        154
    elements 97–98, 131, 134, 138–139, 269
    enclosures 140–141, 141, 264
    exploring 3, 19–20, 26, 43, 130, 154
    exporting 143
    HTML pages 18, 134
    importing 23–24, 142
    properties 18, 29, 43, 133, 138, 271
    standards 20, 133–134, 261, 269
    stylesheets 19, 70, 131–132, 134, 274
    updating automatically 19, 29, 43, 133,
        134, 135, 136, 142–143
    XML 70, 97–98, 132, 134
Metadata tab 12, 18, 70, 130. *See also*
    stylesheets
Minimum bounding rectangle 269. *See also*
    extent
Multiuser geodatabase 64, 65, 269

N

National Spatial Data Infrastructure. *See* NSDI
NSDI 131, 139, 269. *See also* Geospatial
    Data Clearinghouse

O

Object class 270. *See also* tables
Object Linking and Embedding Database
        provider. *See* OLE DB providers
ODBC 79. *See also* OLE DB providers

OGC 270. *See also* OpenGIS
OLE DB providers 64, 79, 270
Online help. *See* Help
Open Database Communication. *See* ODBC
Open GIS Consortium. *See* OGC
OpenGIS 64, 79. *See also* OGC
Origin 40, 202, 204, 270
Overshoot 186, 270
Owner 64, 66

P

Pan 108, 270
Password 77–78, 83, 270
Path 9–10, 48–49, 97–98, 165–166, 270
Path labels 202, 205, 270
PC ARC/INFO coverages 63, 270
Personal geodatabase 12, 64, 65, 100, 270
Pixel 270. *See also* raster: cell
Planar coordinate systems 270–271. *See also*
        projected coordinate systems
Port number 77, 271
Pre-7.0 coverages 63, 271
Precision 184, 186, 264, 271, 274
Preview tab
    defined 12, 14, 104–106, 114
    Geography view 14, 104
    Table view 14, 114
Primary key 40, 41, 202, 206, 271
Projected coordinate systems 74, 191. *See*
        also planar coordinate systems
Projected coordinates 271
Projection 271. *See also* coordinate systems
Projection file 29, 74, 176–177, 182,
        189, 271
Properties
    defined 271
    exploring 14, 48, 93, 99, 154
    sorting 95
    stored in metadata 18, 29, 43, 133, 138
Pyramids 112, 271

Q

Query 61, 70, 131, 202, 272

R

Raster. *See also* data sources
    ArcSDE 66
    bands 66, 105, 272
    catalogs 66, 105
    cell 112, 272
    coordinate systems 177, 189
    dataset 66, 105, 272
    defined 272
    formats 66, 86, 87
    model 272
    previewing 15, 105, 111
    pyramids 112, 271
RDBMS 63, 64, 77–78, 79, 272
Record 272. *See also* row
Reference data 72, 208, 210, 211, 214, 215
Relate 36, 196, 202, 272. *See also*
    relationship classes
Relational database 272. *See also* RDBMS
Relational database management system. *See*
    RDBMS
Relational join 272. *See also* joins
Relationship 36, 64, 202, 272
Relationship classes. *See also* items in
        ArcCatalog
    cardinality 39–40, 41, 202, 203,
        205, 261
    composite 203, 204, 261
    coverage 63, 196, 202–203, 204–206
    creating 39–41, 204–206
    defined 36–37, 63, 202–203, 273
    destination 40, 202, 204, 263
    foreign key 40, 41, 202, 206, 265–266
    geodatabase 64, 65
    key attributes 202, 268
    origin 40, 202, 204, 270